Of Orcas and Men

Drawing by Yukie Adams

Of Orcas and Men

What Killer Whales Can Teach Us

David Neiwert

The Overlook Press
New York • NY

This edition first published in hardcover in the United States in 2015 by
The Overlook Press, Peter Mayer Publishers, Inc.

141 Wooster Street
New York, NY 10012
www.overlookpress.com

For bulk and special sales, please contact sales@overlookny.com,
or write us at the address above.

Cataloging-in-Publication Data is available from the Library of Congress

Book design and type formatting by Bernard Schleifer
Manufactured in the United States of America
ISBN: 978-1-4683-0865-5

FIRST EDITION
1 3 5 7 9 10 8 6 4 2

To Lisa, who makes everything possible

A dolphin appears to be a "who," not a "what"—a being, not an object—with a sophisticated, individual awareness of the world.

—THOMAS I. WHITE, *In Defense of Dolphins*

A human being is part of a whole, called by us the universe. A part limited in time and space. He experiences himself, his thoughts and feelings, as something separate from the rest, a kind of optical delusion of his consciousness. This delusion is a kind of prison for us, restricting us to our personal desires and to affection for a few persons nearest to us. Our task must be to free ourselves from this prison by widening our circle of compassion to embrace all living creatures.

—ALBERT EINSTEIN

Contents

Of Orcas and Men

CHAPTER One

Close Encounters

Y OU WILL OFTEN HEAR IT BEFORE YOU SEE IT; SOMEWHERE IN THE lowering gray fog, there is the loud, almost popping *kooosh* sound. Then you see it: The towering black fin, six feet high, gliding out of the water atop the massive back of a thirty-foot killer whale. Directly toward your little kayak.

You may have prepared for this moment. You may know full well that wild orcas have never been known to attack human beings. You may have observed them around boats and kayaks, and you may already know that these graceful behemoths are in complete control of their whereabouts and always avoid contact. You may have even experienced this previously.

L78, "Solstice," surfaces near my kayak.

It doesn't matter. Your stomach will still disappear into your body, and you will feel puny and powerless. You will know that you are at the mercy of a gigantic predator, the undisputed ruler of the ocean. You may also observe that you had no idea something so large could move so easily and gracefully and quickly through the water toward you.

• • •

Profoundly humbling experiences are good for our souls: those knee-knocking, gut-emptying, jaw-dropping, life-altering moments when you come flat up against the reality that we are each, no matter how big our egos or incomes, insignificant flesh-specks fortunate enough to be alive in this grand universe, those moments such as when we stay up late to see the Milky Way on a summer's night in the Rockies, or stand agape at the edge of the Grand Canyon or an erupting volcano in Hawaii, or watch the birth of our own child. Of all these, there are few as deeply affecting as having an encounter in the wild with one of nature's premier meat-eaters, and of these, none are as profound as having a five-ton killer whale with a towering dorsal fin come looming toward your kayak out of the fog.

It's one thing to see a lion, or a bear, or a shark, or even an orca behind glass at a zoo or aquarium. You can't help but be impressed, of course, not just with the magnificence of the animals, but also with your real gratitude that the glass is there between you. At the same time, there are hardly ever any real interactions with captive animals, at least hardly any that are good for either animal or audience. The barrier keeps everything at a safe remove. This is especially so at the marine parks where orcas are put on display for audiences to "ooh" and "aah" as they perform astonishing stunts of grace and intelligence.

But to encounter such a creature in the wild—well, that's something else.

Over the years, I've been fortunate enough to have had three such encounters in the Rockies, two of them with grizzly bears that happened to be using the same trail on which I was hiking and then biking.

The other was really just a brief but chilling glimpse from the trail I was traversing of a mountain lion running in the other direction. It doesn't matter how big or tough you might think you are or how well you've prepared yourself intellectually for these encounters (frequent visitors to the woods, if they're savvy at all, will always read up on the best techniques for surviving these encounters). Your mind will go blank, your stomach will become a massive black hole, and your body will be seized with a nearly uncontrollable urge to get far from this location as quickly as possible.

This has always felt primeval and visceral, part of our human hard-wiring. Survival of the fittest is often a matter of effective flight, especially for a species such as ours. Man is a top-tier predator with his gadgets and guns at hand, but without them he is just prey like any other, particularly when it comes to the real top-of-the-food-chain predators like sharks and tigers. Before we invented our killing technology, we probably survived by knowing when and where to run.

Fortunately, my wild encounters in the Rockies were utterly harmless. They drove home, however, a lesson not to be forgotten about humans' relative puniness in the grand scheme of the living world. Sure, I could always bring along a good can of bear spray (and tended to thereafter), but the truth is that even that is never a sure thing. Guns are similarly about as useful; one might scare them off, but grizzlies in particular are known to easily absorb or deflect gunshots. If one wants to eat you or just kill you, it probably will.

Killer whales are the planet's only apex predator about whom you can say that the potential for being killed and eaten really doesn't exist. Even though they have been observed countless times over the centuries devouring species ranging from humpback whales to sea lions to moose, there have been only a few recorded attacks on humans by orcas in the wild, and those mostly against boats carrying humans. (This benign relationship with humans cannot be said of orcas in captivity, but that is another part of our story.) And yes, you can tell yourself, as you launch your kayak into waters frequented by killer whales, fully aware—indeed, often eagerly hopeful—that you might encounter them there, that you will not be in danger. However, when the reality comes to pass, and that six-foot fin comes looming toward you out of the gray mist, you remember none of that.

You won't always get the auditory warning. If it is a typical clear summer day in the San Juan Islands, you can see them approaching along with the phalanx of whale-watching boats they attract that time of year. A good watchful eye will still be able to spot them at a distance on a seasonal grey and fogbound day, since even when they are at their most jagged and furious, there is little mistaking the way an orca's dorsal fin erupts from the surface of Haro Strait waters.

Still, more often than not, you will know with utter certainty that you are occupying the territory of killer whales when you hear them breathe, a sound that moves across the surface of the ocean at a remarkable distance, the popping sound of pursed lips exhaling in a whooshing burst, followed by the heavy intake of the big breath the orca takes in. When you hear this sound—and then you spot the misty plume of expelled whale breath and the scythe-like black fin cutting through the water—you become acutely aware that you are now in the water with a very large animal with very large teeth, easily capable of knocking you into the water. If you are an ethical and considerate kayaker, you have already tucked into a cove or kelp bed, well out of the whale's path, but that is only a partial comfort, because no matter what, you are at its mercy.

This, naturally, evokes that primeval response, the blank mind, the emptied insides, the powerful urge to flee. Stuck in a kayak, all you really can do is try to make sure you're out of its path and let the orca do the rest, and no matter how well you may have prepared yourself, your heart will be in your mouth.

There are two exceptions to this rule. The first applies to the scientists and naturalists who see the whales daily in the course of their work, and almost universally for these people, the electricity of the encounters becomes less intense, while the awe and wonder that the whales inspire is replaced after a while with profound respect and deep affection.

The second exception applies to three-year-olds.

• • •

When my daughter Fiona was three, we went out for a paddle on the west side of San Juan Island, where we were camping. My brother's wife, Trish, was in the front of our long two-person kayak, which features a center hatch containing a child seat, complete with a nicely retentive spray skirt. That's where Fiona was seated when we came upon the killer whales.

We really weren't expecting to see any orcas that afternoon. Sightings had been scarce the previous few days, and then a pod of about twelve whales came by the park where we camped that morning. Most of us had watched from shore as they swam past, headed north. The grownups' low expectations notwithstanding, Fiona was still hopeful we would see the whales. The night before, as she had snuggled into her sleeping bag, I had read her a book by a local children's author, Paul Owen Lewis, titled *Davy's Dream*, about a boy with a sailboat who befriends the local orcas by, among other things, singing to them. So when we got in the kayak, despite my warnings not to be disappointed, she felt certain we would see them ourselves that day.

The water was glassy and calm, the day windless, and the currents, which, in the San Juan Islands, can become powerfully riverlike, were mild and easy. About three quarters of a mile south of camp, we rounded a corner from which the park was no longer visible, and almost simultaneously, we came upon the orcas. Actually, they were still quite a ways off, but we knew they were there because the daily flotilla of whale-watching boats that accompany the resident pods from about 10 a.m. to 5 p.m. every day in the summer was forming at the southern end of our view, near the Lime Kiln Point lighthouse. And then we heard them. And then we saw them: tall black fins, heading more or less in our direction.

We were already close to the rock-cliff shoreline, and I tucked the big kayak in a little closer, although I knew it made little difference in whether the whales decided to pay a visit or just trucked on past at cruising speed, as they so often do. That *kooosh* sound, wafting over the half-mile distance between us, announced their presence as well as the fact that we were now in their territory and we were at their mercy. They went where they wanted, at whatever speed suited them. They were large and in charge.

I barely needed to point them out to Fiona; she heard the blows, saw the big fins at the same time as both Trish and I. Still, I talked to her: "Here they come, honey! See them?"

Oh yeah, she saw them, and she began singing to them.

Her favorite movie at the time was the Disney musical version of *The Little Mermaid* (yes, she loved and still loves all things oceanic), which at one point (during key transformation scenes) features a lilting three-note choral melody, and this was what she chose to sing to the approaching orcas. She was relentless, too.

"Ah ah ah . . . Ah ah ah . . . Ah ah ah . . ."

The whales appeared to be in rapid-transit mode as they approached, but now they were slowing down and milling, as if they were hunting the Chinook salmon that are their dietary staple. It took ten minutes or so for them to pass in front of us, but Fiona sang that theme for the entire time. And it was a close pass.

A large male, with one of those six-foot dorsal fins, burst with a *kooosh* out of the water about twenty yards away from us, swimming in a line perpendicular to the boat. We could hear the deep inhale that usually followed. And then he went down and swam away.

"See, Daddy?" Fiona cried. "It worked!"

• • •

We laughed heartily, but I am not entirely certain she wasn't right, because the big male wasn't the only orca who visited us that day. Shortly after that initial encounter, I spotted another orca coming in very close to the rocky shore, behind our kayak. When she submerged, I could see her, clearly diverting in our direction, and I could see her white eyepatch. She was turned on one side and looking up at us. Actually, looking up at Fiona, it seemed to me.

Sure enough, she surfaced shortly, no more than ten yards behind my rudder, and spouted on us. I know people who crave being spouted on, but trust me, it's a mixed blessing. Whale breath reeks of rotting and half-digested fish. Despite all that, it was thrilling and not just a little disturbing.

Was she attracted by Fiona's singing? Or perhaps just by Fiona herself? And if the latter, was it an interspecies matronly response to a young child? Or was it perhaps because she made a potentially interesting lunchtime meal?

I knew, of course, that the latter was beyond unlikely. Recorded attacks on humans by killer whales in the wild could be counted fewer than the fingers of one hand and never with this particular population of orcas. However, that barely salved my conscience. A little later that summer, this same pod of whales was observed "playing" with some of the resident Dall's porpoises—little black-and-white dog-sized cetaceans who can dart through the water at 30 knots and who typically are ignored by these salmon-eating orcas—by taking them underwater and keeping them there. Eventually, the little porpoises would disappear.

There was no doubt that day or on any of the many subsequent summer days Fiona and her mother and I have spent in the occasional company of killer whales that we were at the mercy of those whales. That's a good word for it. Indeed, in their encounters, "mercy" is the word I think most aptly describes the common response of orcas to humans. Unlike their counterparts among apex predators—say, grizzly bears or great white sharks—the encounters inevitably are benign and harmless. However, you can't help but be acutely aware of the whales' forbearance. They could easily bring you to grief, but they choose an intelligent path. They are at best mildly curious about you, or more often they are just supremely unafraid and pay you little mind.

I've been observing these orcas for many years—mostly from shore, but also from a variety of watercraft—and have yet to see a single instance of one of them having even slight contact with a boat (though you know it must happen on occasion, as the propeller-induced notches on some of the whales' fins silently attest). Despite their immense size, these whales are phenomenally graceful in the open sea and utterly in control in their world. This despite humans' best attempts to provoke them.

I've seen an idiot kayaker dart out directly into the path of an approaching male orca in order to experience that "spout" and the ensuing close proximity to a wild whale. Of course, all that happened was that the oncoming whale dove quickly before reaching him and then stayed submerged for another couple of hundred yards, by which time the orca had passed by all of the other kayakers in the man's group who had also come to see whales that day. Cluelessly, the kayaker raised his paddle with both arms and celebrated with a "Yoo hoo!"

I've seen sport fishermen rev up their little twenty-footers and go roaring at twenty knots directly through the middle of a pod of whales who had attracted a crowd of whale-watch boats. Just as they hit the point where the pod had been the thickest, a large male stuck his head straight up into the air in what's called a "spyhop" and then quickly sank back as the little fishing boat whizzed by, only feet away. I'm not sure who would have been damaged more, the fishermen or the whale, but it wouldn't have been pretty.

There's a local legend in the Pacific Northwest about a group of three drunken fishermen who, in their little skiff, approached a pod of resident orcas near shore. They were observed by witnesses inexplica-

A speeding recreational boater in a near-collision with a spyhopping orca.

bly harassing a big male, whose length exceeded that of their boat by ten feet. Finally the fed-up male charged the skiff and, just before hitting it, leapt in a half breach parallel to the boat, sending up a huge wave that washed the three men out of the boat and into the water. Their cries of terror rang in the air as they attempted to scramble back into the skiff, especially since the big male and his black fin kept lingering in the area. Of course, the orca was merciful and let them return untouched to the boat. The soaked drunks scurried back to harbor, and the whales headed wherever they were headed.

Really, you don't want to be messing with a five-ton superpredator, no matter how cuddly he may be made to look at SeaWorld.

• • •

Shamu is the generic name that SeaWorld applies to all of its performing orcas, taken from the original Shamu, a member of the Southern Resident orca community, who was captured as a calf near Tacoma when her mother was harpooned and dragged to Seattle by her captors. She died six years later, but her name lives on in the strange half-life of corporate marketing. In the end, Shamu and her successors, many of them also Southern Residents, made SeaWorld a multi-billion-dollar business. Thanks to the presence of their performing killer whales, the onetime backwater business of marine parks has blossomed into a big-money corporate undertaking of entertainment venues around which families plan their entire vacations.

I know this from personal experience. We took Fiona to Sea World in San Diego a couple of times when she was little—ages one and two, respectively. It was actually a lot of fun.

And you have to give credit where it is due; my little girl was flabbergasted and smitten by the sight of the great orcas, especially when they glided past the glass enclosure where she spent the better part of an hour ooh-ing and aah-ing over them. These parks deserve great credit for providing people the opportunity to actually see, in the flesh, one of these great creatures, but do they truly show orcas as they really are?

Those parks tell a different story about killer whales. They portray

Fiona and Lisa at Sea World.

them as docile and friendly, like super-smart performing dogs. Although imposing and intimidating, they are clearly dominated by their human trainers. The parks claim to give their attendees an "education" and "conservation message" through their shows, although the information they give is often muddled and sometimes downright false. There is a lot of prattle about how much the whales eat and what it's like to train them, but almost nothing about their social lives, their backgrounds, or their population type.

But most of all, you will never, ever hear about the endangered population of killer whales in the Pacific Northwest and most certainly not about the outsize role played by the captive-orca industry in driving those populations to the brink. Yes, mostly as a means of rationalizing the continued captivity of killer whales by theme parks, you may be told that whales face all kinds of survival challenges in the wild and often do not survive them. You will be told the oceans are a scary place (as proven by the difficulties whales face in Puget Sound) and that the parks can provide the whales better food and care than they can get in the wild. You will probably also be told that their orcas live longer in captivity than do whales in the wild.

These are at best gross distortions of reality, and the last is simply a lie. Captivity has been a catastrophe for most killer whales taken from the wild. Study after study has demonstrated that whales in captivity are more than two and a half times more likely to die than whales in the wild. All the care in the world cannot compensate for the stress brought on by placing a large, highly mobile, highly intelligent, and highly social animal with a complex life into a small concrete tank.

Of the 136 orcas taken in captivity from the wild over the years, only 13 still survive. The average lifespan in captivity so far is about eight and a half years. In the wild, the average rises to thirty-one years for males and forty-six for females. Then there is the upper end of the spectrum. In the wild, males will live up to sixty years, and in the Puget Sound, there is a matriarchal female named Granny who is believed to be a hundred years old. Having met Granny up close in my kayak, I can attest that she remains spry and playful. There are several other elderly females in the Southern Resident population. However, we don't know how long orcas will live in captivity yet. We're still finding that out.

Perhaps the most telling number is that, of those fifty-five orcas taken from the Pacific Northwest, only two remain alive today: Corky II, the matriarch of the Sea World orcas in San Diego, taken from Pender Harbor in 1969, and Lolita, also known as Tokitae, the only surviving orca from the horrific Penn Cove captures of 1970, who is alone in a small tank at Miami Seaquarium. Both are estimated to have been born in 1966, making them roughly co-equals as the oldest whales in captivity, although there are some estimates that indicate Lolita is older. Regardless, Corky has been captive longer than any living whale.

On our second visit to Sea World, we bought tickets for Fiona and her mom to go to the exclusive "Dine With Shamu" luncheon, where trainers bring whales up close to the tables where you're noshing and give you a good look. The whale Fiona got to see up close was none other than Corky. I asked her afterward if she was excited to meet Corky up that close. She seemed noncommittal. Her mom told me that she seemed more taken aback than anything. "There was something disturbing about it," she said.

Later that summer, we had our close encounter with orcas in the wild. That evening in camp, I asked Fiona if she had thought about those orcas at Sea World after seeing these in the wild. She said she had. I asked what her thoughts were. She paused for only a moment.

"They should let them go," she said. Even a three-year-old could see it.

• • •

When you first enter the stadium at Miami Seaquarium where the killer whale Lolita does her twice-daily performances, you can stand at the railing and look up close at her enclosure. It is there that you are most overwhelmed by the indelible impression that this fifty-year-old semi-oval concrete tank, with its big concrete island in the middle, is way

Lolita, aka Tokitae, eyes her admirers at the Miami Seaquarium.

too small for a twenty-foot, four-ton orca, especially for four decades' worth of captivity.

Sometimes Lolita herself will come up to you while you stand there. She did so when I arrived early for the second show of the day. I was by myself, and she came up and looked directly at me, then rolled her head around, letting me take photographs, perhaps posing. As with so many resident fish-eating orcas, the abiding impression was of a powerful empathy. After a while, I stopped shooting and just stood there at the railing, watching her watching me. Soon I found myself almost overwhelmed with feelings of affection, of caring, for this whale. She just looked back, but when a whale looks back, it's not the same as when, say, a dog or a cat looks back.

A Brazilian man sidled up next to me, and we began chatting about Lolita. I told him she was about forty-seven years old and had been in that very tank we were looking at for nearly forty-four years. He did a double take. "Really? Isn't this tiny for an animal this big?"

I smiled. "Yes," I said. "That's a problem." I said no more. The docents were listening, and I didn't want to be thrown out.

A killer whale petroglyph carved by Makah tribal watchmen, at Wedding Rock on the coastal Olympic Peninsula.

CHAPTER Two

The People Under the Sea

T HE ORCAS WERE WELL KNOWN TO THE NATIVE AMERICANS WHO occupied the Northwest Coast. They shared these waters and the fish that swam in them. Harming orcas was universally taboo for these people. While many of these tribes avidly hunted humpback and fin whales, it was considered bad medicine to injure or kill a "blackfish," who at the least was a harbinger of plentiful fish and often much more. It was widely believed that if one killed an orca, its family would wreak vengeance on you and your kinsmen the next time you took to the water. Orcas were not mere beasts, but the people who lived under the sea.

Most of these tribesmen believed in a realm parallel to ours occupied by spirits, many of whom were people who lived in spirit villages rather resembling the natives' own large cedar dwellings. Foremost among these spirits were the killer whale people, whose powers were immense and far-reaching. This was why so many tribes claimed the killer whales as their spiritual totem and carved the creature's likeness into their totem poles and family crests. At times, the killer whale would meld with other creatures, especially wolves; they were often described as being the wolves' seagoing counterparts. One of the most fearsome mythical creatures, the Wasgo, was half wolf and half killer whale. The orcas' only predator was the great Thunderbird, which was large enough to cart off orcas and great whales and other creatures; there appears to have been only one of those at any given time.

According to the myths of the Kwakwaka'wakw tribes of northern Vancouver Island (also known as the Kwakiutl), the first men were killer whales who came to shore, transformed into land creatures, and then forgot to go back. Some of their tribal elders claimed direct descent from killer whale tribes. Their word for killer whale, *Max'inuxw*, means "the ones who hunt." In this mythology, the killer whale is the lord of the underwater realm. His house can be reached by four days' journey out to the open sea, and his village is at the head of a long narrow inlet. The dolphins are his warriors, and he sends out his messenger and slave, the sea lion, with the king's four quartz, pointed wedges, mystical devices that can never be blunted even when used to split stone.

"Our people have a great respect for the whales, because our belief is that they are our ancestors," says Andrea Cranmer, a storyteller with the Kwakwaka'wakw tribe in Alert Bay, British Columbia (B.C.), near the northern end of Johnstone Strait. "They come and visit us, usually, at the beginning of a potlatch, if the family is descended from the whales, the *Max'inuxw*, the killer whale. Also, it's a dance in the potlatch. Living on the island, they don't show up every day, but when they do show up, we're like tourists, too. We're excited to see the ancestors. And when you see the dorsal fins, when the big dorsal fin comes out, and we know how old they are? It is just exciting to see, because it's so massive. And there's usually a pod—there's usually not just one whale. So you see a whole family of whales, and it just makes you all excited."

For nearly all Northwest tribesmen, including the Haida, the killer whale embodied the spiritual and physical power of the ocean. In various myths, humans come into contact with them and emerge with vast spiritual powers themselves.

One legend tells of a boy who, alone at sea in a storm, is thrown from his small canoe and swept underwater. There, under a strange sky, he encounters a village occupied by even stranger men, huge and black in the face. The villagers welcome him as the son of a chief and invite him to partake of their fire and the salmon on which they are feasting. Afterward, they all dance together and teach one another their tribal dances. As the festivities wind down, the strange chief tells the boy that he knows the boy wants to return to his home and commands him to close his eyes and take hold of the chief's staff. The boy does so, eyes tightly shut, and finds himself grasping the great fin of a killer whale and rising upward. He awakens on the pebbly shore in front of his village and is told by his relieved mother that he has been gone a year, not the single night he experienced. He goes on to teach his village the dances taught to him by the killer whale people under the sea and becomes in time a chief and shaman of great power.

The killer whale's spiritual potency is symbolic of the power of blending human intelligence with the forces of nature. The orca's spirit is revered by Northwest tribes because it is seen as a friend to all of mankind. But *Max'inuxw* is also a fearsome thing to behold, its awful vengeance only a swish of its flukes and a flash of its toothy jaws away. *Max'inuxw* is not Shamu.

. . .

In ancient times, the Kwakwaka'wakw say, men and killer whales could talk to each other. Unfortunately, they were also at war. The myths do not say why, but for many years, orcas and men hunted and killed each other. One of the dances performed at the lodges tells the story of how the depredations of the orcas became so great that the people prayed to the spirit of the mighty Thunderbird to save them. In the dance, the Thunderbird and the *Max'inuxw* do fierce battle until finally the Thunderbird strikes a mighty blow, and the Killer Whale lies dead.

As the storytellers say, it was a time of pain and fear for both men and whales, but it did come to an end. It seems a boy from the Ma'amtagila tribe, whose territory included the land we now call Robson Bight, somehow befriended the killer whales there. So he decided the killing and fear had to cease. He began telling this to killer whales he would meet. One day he was out hunting, and the chief of the killer whales came to him. "Let us talk. Our losses are becoming great. We are concerned," he said.

The boy replied, "We too are suffering. We are concerned, too."

The *Max'inuxw* chief told him, "You must speak with your people. Tell them we must end this heartache. Tell them we must no longer hurt each other."

The boy promised he would do as the chief asked. "I want us to be friends," he said.

So the boy returned to his village at Itsikan and went before his chiefs at the great assembly. They heard him tell of the agreement proposed by the *Max'inuxw*. They huddled among themselves, and then called the boy before them.

"You must return to the *Max'inuxw* with word that we have agreed to this. And we will agree to it for all time. Take it back to the whales and tell them."

So the boy did as the chiefs asked. The first peace treaty between killer whales and man was made, and ever since that time, men and the orcas have lived together without killing each other. The *Max'inuxw* bring salmon to the tribes, and the tribes honor their ancestors, the *Max'inuxw*.

"Today, the Kwakwaka'kwa still honor this treaty," their storytellers say. "There are many stories of how *Max'inuxw*, the killer whale, has helped us. They have saved people from drowning; they have shown us where to find food in times of famine; and we know the whales to be our ancestors' spirits. And we have lived in harmony with *Max'inuxw* ever since the boy befriended them."

• • •

There are still people who claim to be able to speak with killer whales—psychically, that is. One of these lives in the San Juan Islands and has written a book about the subject, describing telepathic conversations she has had, from the comfort of an apartment in San Francisco, with Granny, the matriarch of the Southern Residents' J pod. Oddly enough, Granny's psychic descriptions of life in the pod happened to coincide almost perfectly with scientists' previous descriptions of the daily life of killer whales, with a few juicy tidbits thrown in for good measure. Whale scientist Ken Balcomb says the believers in psychic contact call him up at his Center for Whale Research from time to time. "I've had people calling me up from Montreal, saying, 'I've just had this feeling about Granny, and they just sent me this telepathic message,' and I was supposed to go out and check on them right away," he recalls with a chuckle. His answer:

"Um, did she tell you where she was? Because it's January, and I have no idea."

Attributing mystical values to cetaceans, and particularly to dolphins, has become a cottage industry around the globe, running the gamut from 3-D oil paintings and kitschy sculpture to abusive "swim with the dolphins" programs that purport to have the ability to heal the sick. Along the way, the public has adopted a number of myths about dolphins that either aren't true or are distortions of things we simply don't know with any certainty. It is almost an article of faith, for example, that dolphins have kind and cooperative dispositions, are models of non-aggressive behavior, and have a special attachment to humans, none of which is true at all.

The most appalling permutation of this mentality has been the proliferation of a variety of facilities that offer the opportunity for people to experience what it's like to be in the water with a dolphin and that often claim all kinds of marvelous medical value from what's called "dolphin assisted therapy" (DAT). The mortality rate of dolphins in these businesses is appallingly high, and it's not surprising that many people discover that, once in the water with a dolphin, the dolphin rarely actually wants to be in the water with them. A number of people have been bitten and bashed by their cetacean "friends" in these facilities.

It's all fakery. As dolphin scientist and ethicist Lori Marino observes, "The worst of it, perhaps, is that there is absolutely no evidence for DAT's therapeutic effectiveness. At best, there might be short-term gains attributable to the feel-good effects of being in a novel environment and the placebo boost of having positive expectations. Nothing more. Any apparent improvement in children with autism, people with depression, and others is as much an illusion as the 'smile' of the dolphin." Marino disparages these programs as "a whole bucket of wrong."

As the largest dolphin, orcas have been subject to similar mythologizing and mystical deification. A fair amount of it is fuzzy-minded silliness bordering on pure charlatanism, such as the psychics and their "revelations" about orca life. Some of this is embodied in the fuzzy, kid-friendly image of orcas peddled at marine parks around the globe, all of which erases the whale's essence as both a predator and a social animal.

However, nearly everyone who is around orcas for much time at all—either in the wild or in captivity—will attest that they have an uncanny knack for the seemingly coincidental; they do things, or show up at times, that suggest that the animals are somehow reading people's minds and messing with them, like the big male orca I paddled with in Johnstone Strait, who mere moments after I had mused to myself about how undemonstrative the Northern Residents appeared to be, breached only twenty yards away from me, as if to say: "I'll show you demonstrative!"

One of the better known examples of this phenomenon occurred in late October 2013. A Washington state ferry transporting a precious collection of Native American artifacts was forced to stop when it was surrounded by a pod of cavorting orcas. The Bainbridge Island ferry runs every half hour or so, and the orcas that day happened to pick the one boat that was transporting rare tribal artifacts from Seattle's Burke Museum, where they had been kept for decades, back to the western side of Puget Sound, where they had originated, to be housed in a brand new museum built by the Suquamish tribe, Chief Seattle's ancestral nation. The tribal chairman, Leonard Forsman, noted that his tribe's myths also celebrated the killer whale and viewed them as ancestors.

"They were pretty happily splashing around, flipping their tails in the water," he said. "We believe they were welcoming the artifacts home as they made their way back from Seattle, back to the reservation."

Indeed, pods of killer whales had been seen throughout the fall at various locales around southern Puget Sound. Whale watchers said they believed the orcas were feeding on a large run of chum salmon.

"We believe the orcas took a little break from their fishing to swim by the ferry, to basically put a blessing on what we were doing on that day," Forsman said.

Most of the scientists who have worked with killer whales in the wild have stories about the odd coincidences that keep popping up around them. Ken Balcomb still laughs about how the whales always

Just when you're looking the other way ...

seemed to know, in the days before digital photography when he was gathering photo-identification information, when he had reached the end of his roll of film. "That's when they would always do the really spectacular stuff, and you'd never catch it on film," he says with a rueful smile.

It's true of captive whales and their trainers, too. When Alexandra Morton worked with the captive Northern Residents Corky and Orky at Marineland of the Pacific (before that facility was purchased by Sea-World and the two whales transferred to San Diego), she had a funny experience along these lines. She had just returned from a trip to Vancouver Island to meet with Mike Bigg and Paul Spong and see the whales' home pod and waters. She was discussing with one of their trainers, in Corky's presence at the stadium, how she went about introducing new routines for the whales. The trainer asked Alex to suggest a new behavior that could be performed by the whales.

Morton thought back to what she had seen in Johnstone Strait and suggested they try something natural to the orcas like a pectoral slap, when the whale glides along the water's surface on its side and slaps its pectoral fin on the water, something they had never seen either Corky or Orky doing. The trainer thought this sounded like a good idea and indicated she would try it. As the trainer walked out of the stadium, Corky turned to Alex and did a pectoral slap. First one, then another. Pretty soon she was just swimming around the pool on her side, slapping happily away. Alex called the trainer back so she could see it for herself.

"That's whales for you," the trainer told Morton with a laugh. "They can read your mind. We trainers see this stuff all the time."

• • •

It seems as though nearly everyone who spends time around orcas has a tantalizing anecdote—or several—to tell. It would probably be possible to collect them all and write an amusing book about that. However, in the end, all they really are is anecdotes, hinting, suggesting, teasing us with the possibility that there is an intelligence, if not outright psychic powers, beyond our understanding at work with killer whales. At the same time, nearly everyone who experiences these incidents will acknowledge that in fact each could be pure coincidence, combined with an eagerness on our part to read motive into the random. We simply don't know. Moreover, the kind of phenomenon we're describing barely fits into any scientific category and may well be beyond anything scientifically testable. In the end, it really tells us nothing at all about the nature of these creatures or their intelligence and may obscure reality with a lot of hokum.

The question that hangs at the center of all this speculation is a fairly simple one: *Just how intelligent are orcas, really?* The answer, however, is anything but simple, other than to say that we really don't know—and a lot of the answer depends on how you define "intelligent" anyway.

There already has been a considerable backlash in the scientific community directed toward much of the woolly material that has been published about dolphins. Justin Gregg, an accomplished dolphin scientist, has led much of the debunking, pointing out the limits of much of the more remarkable findings regarding dolphins, including Diana Reiss's and Lori Marino's work establishing that dolphins are capable of self-recognition in a mirror, considered one of the markers of a high level of intelligence. (There have been similar indications of self-recognition in captive orcas.) As Gregg notes, these studies are severely limited by their small data samples, and it is not really clear how much intelligence mirror self-recognition denotes.

Moreover, as Gregg elucidates at length in his book *Are Dolphins Really Smart?*, much of the mythology about dolphin behavior is unadulterated nonsense. They are not particularly peace-loving. In fact they often bash and bite each other and the humans with whom they come into contact. Bottlenose dolphins do not, in fact, display superior cognitive skills in certain key tests relative to a number of other animals. Also, while the size and complexity of the dolphin brain is impressive, it is also different enough from the primate brain in fundamental structure that it is impossible to draw any scientifically valid conclusions based on the limited amount that we actually know about brain physiology.

The story is much the same, but much more complex, when it comes to killer whales. There has been relatively little extant testing for behavioral intelligence in orcas, in part because the number of captive orcas is so small and the demand for data slight. However, a study published in 2013 did test captive orcas for imitative learning and found they had exceptional capacities in this regard and that some of these skills might help account for the social behavior of killer whales in the wild, especially group-specific traditions that are handed down from one generation of orcas to another.

Those same traditions, however, indicate some of the limitations of the intelligence of killer whales. As far as having a healthy gene pool ripe with diversity, for instance, it does not make sense for killer whales to limit their mating activities to other killer whales within their larger community, which is how, as genetic scientists have recently determined, an array of orca "ecotypes" around the globe behave. Even though it may have made sense during an epoch of abundant prey, from a long-term biological perspective, orcas' ultra-conservative social structure carries real negative consequences for their survival.

Similarly, the cultural rigidity that expresses itself in their unusual selectivity when it comes to prey really limits their food choices and their abilities to feed themselves. As scientist John K.B. Ford has found, both the Southern and Northern Residents almost exclusively dine on Chinook salmon when they are present in the Salish Sea waters, which is much of the year; in the periods when Chinook are in decline, they will feed opportunistically on chum salmon but instead tend to travel to areas of the open sea where Chinook can be found at these times. Were they not so rigid about what they eat, they would find plenty more food, including abundant runs of pinks and sockeye salmon, as well as a bounty of non-salmonid species, such as lingcod and rockfish.

A cold-eyed examination of one of the more tantalizing aspects of killer-whale intelligence, their communications and especially the dialects of their pod-specific calls, does not enhance the case for their intelligence, either. Once studied, most orca calls in the wild consist of a very limited vocabulary of about 40 sounds used by each whale, a number of which are reiterated monotonously. If each of these calls symbolizes a specific thing to orcas, in the way that we understand languages to function, then they appear mostly to be having conversations that run something like this: "Hey." "Hey." "Hey." "Hey." "I'm here." "I'm here." "Over here." "Hey." "Here." "Wait." "Fish." "Hey." "Hey."

Some of the scientific critics of the claims of dolphin intelligence insist that dolphins really are no more intelligent than dogs, and there are some who think the same of killer whales. Jim Borrowman, a long-time whale-watching guide from Telegraph Cove, B.C., told me that after years of being around them, he believed they were only about as smart as dogs or wolves, although he had real respect for the intelligence of the latter.

"They're plenty smart," he told me, "but not *that* smart."

• • •

So how intelligent are orcas? Well, how do you define intelligence?

Here's a thought experiment: suppose the scientists here on Planet Earth who are engaged in searching for signs of extra-terrestrial life were to find evidence (via, say, an interstellar probe on some distant aquatic planet) of a life form that turned out to be identical to killer whales or some other dolphin, with all the signs of intelligence, but with a similar communication barrier. Would it still qualify as an intelligent life form?

I asked Seth Shostak, the senior astronomer at the SETI Institute, whose chief work entails hunting down the possibility of intelligent life on other planets (SETI is an acronym for Search for Extraterrestrial Intelligence). Shostak is also a well-known skeptic and critic of the pseudo-science that surrounds this field. Unsurprisingly, he quickly shot down the fanciful question for its impossible qualities, because, of course, as smart as these creatures might be, they don't create technologies capable of flagging down intelligences from other planets.

"In terms of what would happen if we were to find dolphins on another world, the simple answer to that is, we won't," he told me. "I mean, dolphins do not build radio transmitters. So you can say they're intelligent, but not according to our definition. If you can't solder together a transmitter, we don't hear from them."

The subject of dolphin intelligence, however, has special resonance for the scientists at SETI since there was a good deal of research into dolphin communications in its earliest days, due largely to the presence of Dr. John Lilly, the pioneering cetacean researcher, in the gatherings of scientists concerned about finding extraterrestrial intelligence. One of the earliest of these, in November 1960, was dubbed "the Order of the Dolphins," and it included such astronomical luminaries as Carl Sagan, Frank Drake, and Otto Struve.

"There was hope in the early days that if we could learn how to communicate with the dolphins, that would give us some indication we might be able to communicate with E.T.," Shostak said.

Of particular note in recent years has been SETI researcher Laurance Doyle's work examining dolphin sounds. Doyle applies a branch of mathematics called information theory, the study of the structure and relationships of information, to analyze radio signals coming from space, in the hopes of detecting one of those signals that Shostak described. He and other scientists, notably Cal-Davis's Brenda McCowan, have tried applying the same theory to noises from bottlenose dolphins. They found that dolphins, in comparison to other animals particularly, have a language complexity comparable to humans. McCowan found, for example, that adult dolphins send information when they whistle, while infant dolphins do not; much like human babies, they only babble in comparison. Similar studies of humpback whale sounds establish that the strange songs they sing feature their own kind of syntax and have many of the basic features of a language.

The problem of the communication barrier is not, evidently, a matter of cetacean stupidity. Indeed, although bottlenose dolphins do not have the physical ability to imitate human speech, scientists at the Dolphin Institute in Hawaii were able to teach them over a hundred human words. The dolphins were also able to comprehend entire sentences, including syntactic nuances.

Indeed, Sagan once famously observed, "It is of interest to note that while some dolphins are reported to have learned English—up to 50 words used in correct context—no human being has been reported to have learned dolphinese."

There are plenty of skeptics, of course, Justin Gregg among them. "Most scientists, especially cognitive scientists, don't think that dolphins have what linguists would define as language," he said. "They have referential signaling, which a lot of animals do—squirrels and chickens can actually do that, and monkeys—and they have names for each other. But you can't then say they have a language because human words can do so much more."

At the same time, we are only beginning to get a glimpse into what it is that orca and dolphin communications really can do. "Dolphins have exquisite sound and they have a lot of places they could potentially encode information—we just haven't looked adequately yet," says dolphin researcher Denise Herzing. Examinations of the echolocation signals emitted by killer whales, for example, have revealed that they are comprised of extremely dense information packets, suggesting that the information they obtain in the pingbacks from those signals is very rich indeed.

Moreover, those huge brains, for all those similarities to human brains, are structurally quite different in a fundamental way: the nerve fibers that comprise the bulk of dolphin and orca brains are built for transmitting sonic information, in contrast to the brains of most land mammals, including humans, which are dominated by visual-cortex fibers. In other words, their brains are particularly geared for processing sound information.

This makes a great deal of evolutionary sense, considering that the liquid medium in which orcas and other cetaceans dwell is a tremendous transmitter of sound waves. Light, on the other hand, has problems in water; even the clearest of waters still only allows about a hundred yards of visual acuity for the best of eyes on the brightest and calmest of days, and most cetaceans have excellent vision. However, using their sonar capacities, orcas can "see" underwater for hundreds of yards, perhaps even up to a mile or more, expanding their "visual" realm by leaps and bounds.

Yet if these sonic capacities are simultaneously powerful and subtle, then it seems likely that there is more than greets the human ear to the sounds they use to communicate with each other. If their brains are structurally and fundamentally different, then our criteria for "intelligence" will almost certainly be different than theirs, too. It is just as possible they see us as acoustically obtuse and primitive—since, in comparison to them, we probably are.

For Seth Shostak, our thought experiment about extraterrestrial orcas was at least a good way to provide perspective on our own species and its presumptive claim as the planet's only intelligence. "I think there's a general consensus among people who study this that humans are definitely not the only intelligent species on the planet," he said.

• • •

Most of what we define as "intelligence" involves the ability to form, recognize, and then manipulate abstract ideas. These ideas, particularly the ones for which we test when we want to measure intelligence and which involve measuring, investigating, and conceptualizing the world around us, are all derived from the limited number of senses of sight, hearing, touch, taste, and smell. What we call "intelligence" fits within what we know of those realms. It is a decidedly human-based test.

Orcas, too, have a limited number of senses. Their ability to see is about equal that of humans. Their sense of touch is exquisite and may be more important than we know. They have no sense of smell at all, but they do have a well-developed sense of taste roughly equal to ours.

However, it is in the realm of hearing that killer whales' senses reach another dimension entirely. They not only can perceive the world by the simple reception of sound, as land mammals can, but they are also capable of making sounds that reflect back to them and that, thanks to huge brains capable of translating all this information, enable them to not only see the shape and nature and inhabitants of their world, but to see *inside* them. That is a kind of intelligence that is simply beyond our ability to fully comprehend, let alone measure.

At some point, the breadth of a species' perception (that is, how many different kinds of data it receives from varying sources) and its depth of perception, the level of penetration of reality that its senses provide, should both factor into our assessment of its intelligence. If those are our criteria, then killer whales are definitively, and undeniably, more intelligent than human beings, because their echolocation sense provides both greater breadth and superior depth.

As Marino puts it: "Orcas may not be very intelligent humans, but humans are really stupid orcas."

That, in fact, is the root of the problem. Even as we determinedly avoid anthropomorphizing these creatures, we almost reflexively apply a patently anthropocentric definition of intelligence, one involving language and its use. This is a definition that almost automatically places humans atop the heap, since our wired-in instinct for language is arguably one of our greatest evolutionary advantages.

Dolphin scientist Thomas White (along with others) has proposed an alternative approach to defining intelligence, one that is "species-specific": "The challenges that need to be met simply to stay alive are significantly different on the land and in the water. . . . [W]e need to be careful in making straightforward comparisons between human and dolphin intelligence. It may be like comparing apples and oranges."

Instead, White argues, we should "think about intelligence simply as the intellectual and emotional abilities that make it possible for both a species in general and its individual members to survive in their environment and to solve the problems and overcome the challenges that life throws at them."

Lori Marino points to the killer whales' echolocation as evidence of a different level of intelligence: "I would say that having this sixth sense means that we're getting only slivers of who they are from what overlap there is," she says, noting that most attempts to assess their intelligence have tended to focus on their communications and their "language," if that's what they possess. "Whatever we can recognize in them that we have in ourselves, we throw that into the bag we call intelligence, and we say, 'Well, at least we can recognize this in them and that in them,' and so forth. But you know, that's just a small, narrow sliver of who they are."

• • •

Even if you set aside the questions about their intelligence, it's certainly not a stretch to conceive of orcas as the oceanic counterpart to human beings. They are, in fact, not only the top predator in the ocean—nothing preys on killer whales, while they are known to consume nearly everything oceangoing, including large sharks, as well as sea birds and a stray moose or two—but they are its most successful species. *Orcinus orca* is known to inhabit every ocean on the planet, and while its numbers are not overwhelming, only a handful of populations of the species are endangered.

It is true that, unlike human beings, orcas are incapable of building great technologies or great civilizations. Yet it is also worth keeping in mind that part of their success as a species is reflected in their great longevity. While humans have only been around for about 200,000 years as a species, killer whales have been the supreme creature in the ocean for about six million years, roughly thirty times longer. It could be they have something to teach us about staying atop the food chain.

There is plenty of evidence that should give anyone pause, moreover, before dismissing the possibility of high intelligence in killer whales. The first is that big, complex brain. It is the second-largest brain in the animal kingdom, second only to that of a sperm whale and only slightly larger than an elephant's brain. However, the physical brain only tells us so much in general, and that is especially true if the issue is sheer mass. Like most such measures relating the brain's physical structure to intelligence, it is at best a sort of broad indicator; the largest brains indeed belong to some of the more intelligent animals, cetaceans (both toothed and baleen), elephants, and primates. Within that range there is relatively little correlation with the differences in testable intelligence. Cows, for instance, have relatively large brains, too, but test poorly on a number of intelligence fronts.

In the end, what matters with all these measures is how well they actually predict behaviorally measurable intelligence. It is what they do, not how they are built, that is the bottom line. For years, the intelligence quotient (IQ) was considered the basic means of measuring how smart people and animals are, but over time it has become clear that there are underlying biases built into such tests. In more recent years, it has become more common to test for the "g factor," the general intelligence factor, which summarizes correlations between different kinds of cognition and is considered a more reliable measure of smarts.

Relating measurable intelligence to the physical brain is tricky. The measure that is most commonly referred to is the encephalization quotient (EQ), which measures the size of the brain compared to its body-mass-related expected size (a more refined measure than raw-body-mass-to-brain size) and also happens to favor humans significantly, since they come out on top of most such scales by a good margin.

There are a couple of measures of brain structure that do, however, correlate with cognition. The first of these is cortical thickness, a simple measure of the cerebral cortex, the part of the brain most associated with consciousness, memory, attention, language, and thought. On this scale, cetaceans also rate highly, but not as well as primates and particularly not as well as humans. In cetaceans, the cortex is relatively thin, especially compared to humans, however, cetaceans' cortices are also structured quite differently.

This is where the second measure, gyrification, the amount of wrinkling and folding in the cortex, comes in. What gyrification does

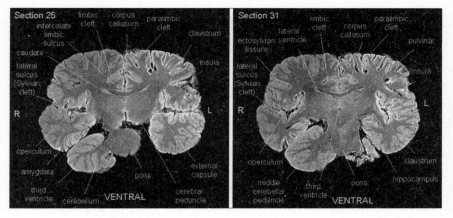

Cross-section of a killer whale's brain, showing the extreme amounts of folding, or gyrification, in its cerebral cortex, and an additional paralimbic lobe, absent in humans and other land mammals.

REPRODUCED BY PERMISSION FROM LORI MARINO, ET AL., "NEUROANATOMY OF THE KILLER WHALE (ORCINUS ORCA) FROM MAGNETIC RESONANCE IMAGES," *The Anatomical Record*, (NEW YORK: JOHN WILEY AND SONS, 2004), FIGURE 3.

for a brain is increase the amount of total cortical nerve tissue dedicated to processing information, making brains with more folds and wrinkles more capable of handling more data and processing it faster. On that scale, cetaceans, especially whales, are significantly ahead of their land-mammal counterparts; the scientists who examine their brains are often astonished at just how heavily folded these brains are. The gyrencephaly index (GI) for humans is about 2.2; for bottlenose dolphins, it is 5.62. The honor of having the most gyrified brain on the planet, in fact, belongs to *Orcinus orca*, which comes in with a GI of 5.70.

What are orcas and dolphins doing with all that brain power? Considering the predominance of auditory nerve fibers in their brains, it is a reasonable guess that they are using it to process information from their incredible hearing faculties. This correlates with the reality that these creatures possess a real sixth sense that we lack, echolocation. In other words, they probably are using those big brains to see underwater, to perceive the world in a way that we never can.

Lori Marino observes that "orcas are the most acoustically sophisticated animals on the planet," and much of that is related to their brain, which like that of most cetaceans has a different architecture than that of land mammals. "It's not a completely alien brain, because they're mammals, and all mammal brains are the same. Especially under the cortex, a brain is a brain is a brain," says Marino. "Just as we have linguistic elaboration, a lot of things that help people conceptualize, they have their own cutting-edge elaborations. I think it would take the form of mental representation for their echolocation, and also mental processing in the form of acoustics. We know, for example, that when they echolocate they get a mental picture. So it's not entirely acoustic. They can go back and forth between vision and echolocation very well. So there's a certain degree of convergence in integration that happens in the brain that goes beyond just the acoustic level."

Indeed, there are some striking differences between orca and land-mammal brains. Scientists point out that cetacean brains have a number of other structures apparently devoted to cognition that are absent in the brains of land mammals. There is very little of the orca brain devoted to the sense of smell, and perhaps more important, the lobe of the brain that ties its two halves together, known as the corpus callosum, is very small in orcas.

At the same time, orca brains have something that humans and land mammals don't, a highly developed set of brain lobes called the paralimbic system. Scientists aren't sure what it does, but they speculate that it replaces the function of some other under-developed lobes in their brains that, in land mammals, are linked to spatial memory and navigation, as well as the brain-tying functions of the corpus callosum, or it may enable some brain function we can't even envision because we lack it.

"We don't know what it does," says Marino. "We can only infer. But what we do know is that this is an area that is not identifiably elaborated at all in humans and many other mammals. So, there's something about those cetacean brains that required them to develop something, probably having to do with processing emotions in some other way, that caused this lobe to elaborate. We don't know.

"It is a very mysterious part, probably the most compelling part of the brain of orcas and dolphins. Because again, we're not used to other animals having things that we don't have, in our brain."

The scientists also found that orca brains have a highly developed amygdala, the part of the brain that's associated with emotional learning and long-term memories.

However, what catches the attention of nearly every brain scientist is the killer whale's insular cortex, aka the insula, one that, according to Marino, is "ridiculously elaborated." Thus the highest GI on the planet.

"And the insula has some very interesting functions in the mammal brain, including not only social cognition, but things that have to do with awareness, self-awareness," she adds. "For instance, the insula is abnormal in people with obsessive-compulsive disorders. So there are things about the insula that have to do with attention, focus, thinking—it's a very, very interesting part of the brain."

At present, we can only surmise what all that brain power is being directed toward. Echolocation, a genuine sixth sense, is almost certainly part of the picture, and the acoustic intelligence orcas obtain from it opens up a dimension of knowledge completely unknown to humans.

And the orcas' amplified empathy? We can only blindly guess what that means. However, the longtime observations of many scientists and trainers that orcas exhibit complex emotions of a wide range and have a powerful, abiding empathy for each other and for humans as well is supported by the physiological evidence, too.

• • •

Orcas certainly exhibit plenty of parallels to humans. We know that killer whales (those in captivity, as well as those in the wild) have massive brains with extraordinarily keen intelligence that they display in many of their interactions with us. We also know that they engage in complex communications with each other in the calls, whistles, and clicks that they use when in proximity to each other in the wild and possibly in their echolocation signals, which enable them to see great distances underwater.

There are other, deeper similarities: Orcas are highly social animals and generally live in tight-knit families. They live about as long as humans (the females are known to live well into their eighties, while the males generally live into their fifties). They display ritual behavior, including "greeting ceremonies" that occur when large resident pods encounter each other after long seasons of hunting in separate waters. Food-sharing behavior is common.

There are cultures, which may extend even into biological differences, displayed by killer whales who occupy the same waters. These salmon-eating resident killer whales with whom we were kayaking are the most commonly sighted orcas in the San Juan Islands, but there are also pods of so-called "transient" orcas (also known as Bigg's killer whales, after the late Canadian researcher, Michael Bigg, who first identified them) who come trolling through as well.

These whales do not eat much fish, but prefer instead to munch on seals, sea lions, and various other marine mammals, including Dall's porpoises and Pacific white-sided dolphins. They have even been observed taking down moose that were swimming, as they are known to do, between islands in the Northwest's abundant marine archipelagoes. Unlike resident orcas, transients are not very gregarious in the water, preferring to hunt in silence, although they will vocalize avidly after a successful kill. Their stealth also makes them rather scary to encounter in a small boat in the water.

Resident killer whales and transient killer whales do not mingle. They hunt in different zones, the transients preferring the shallow bays and rocky haulouts where seals can be found, the residents preferring the open straits where the salmon are running. They do not encounter each other often. When they do, the resident orcas—who typically outnumber the transients, sometimes by significant amounts—aggressively chase the transients away. This may be why genetic tests indicate that transients and residents have not interbred in perhaps as long as 700,000 years, longer than humans have even been around.

The communications may be the most remarkable indicator of their intelligence, as well as their cultural proclivities. A Russian scientist studying the sound organization of the communications of various dolphin species, including orcas, observed that "both for humans and dolphins the sound organization is nearly the same in terms of complexity. On the whole, there are lots of noticeable parallels between the two species: *Homo sapiens* and *Orcinus orca*."

I purchased a hydrophone (a microphone for listening to underwater sound) several years ago so my family and I could listen in on the killer whales when they went past our kayaks; most of the time, in this way, we could hear them even from considerable distances. The sounds that we have heard over the years have always seemingly involved members of the pods communicating information to each other; the call and response interactions were clear, and the variety of the sounds made it obvious that this was not just random noisemaking.

Moreover, the sounds we were hearing were part of these whales' distinctive dialects, according to the researchers who study them. Southern Resident killer whales have a unique set of calls they use when talking to each other. These calls are similar to, but quite distinct from, calls used by Northern Residents from Vancouver Island and not even remotely like calls used by transient killer whales or killer whales in other populations who have been studied, including those in Iceland, Antarctica, and New Zealand. Moreover, researchers discovered that the orcas use discrete calls that identify them not just by their clan (that is, Northern or Southern Residents) but by their individual pods as well.

What are they communicating? What are they saying to each other? If someone could crack that code, it might breach the barrier between humans and animals.

"This question of recognizing intelligence and self-awareness and personhood in other animals is a difficult one, because we are limited by whatever we can understand and experience," says Lori Marino. "The question is: Is there some way of ever getting around that? I don't think we know." Nor, says Marino, are we likely to. "We have been trying to find this Rosetta Stone for communications with orcas and dolphins, and everything we do applies some human-centered criteria—only because that's really all we know, it's all we can do. When, in fact, if the principles they use for communication are outside of our own type of system, it's going to be very difficult if not impossible for us to understand with what we would call 'the code.' And so that's going to be very, very, very hard."

I AM PRETTY CERTAIN THAT THE FEMALE ORCA DRIFTING TOWARD MY kayak is not very happy that I am there. For that matter, I am not all that happy, either. As seems to happen with killer whales, we all kind of surprised each other, and now I am stuck.

I hadn't intended this. The afternoon at camp on Kaikash Creek had grown quiet, and I had decided to get back on the water and see what I could see. My instincts were telling me the whales were getting near. They were right, too. No sooner had I put my little blue craft into the deep glasslike Johnstone Strait waters than a pod of orcas came tooling around the corner to the north of camp.

An A-pod female wearing a water bubble.

At that moment, I stop where I am, about twenty feet out from shore, and watch them approach, dropping my hydrophone into the water and pulling out my telephoto camera. Most of the pod, about twelve in all, stay out somewhat mid-channel. A whale-watching boat pulls up out further in the channel, and the tourists all come out on deck to see.

However, there are a calf and its female minder—probably its mother, although possibly one of its "aunties" acting as a sitter—coming in closer to shore. The calf appears to be foraging, possibly chasing one of the many thousands of pink salmon that have been flocking through this wild strait, tucked along the inside shore of northern Vancouver Island, for most of this week. So many salmon have been leaping out of the water as I've paddled through here that I've thought for sure one was going to land on my boat and give me a free dinner.

I can't really tell what the calf is doing because it is remaining submerged for long periods, even though it seems it is in close to the rocky shore. I paddle back as far as I can, until I'm in danger of hitting some of those rocks, and then stop, knowing the orcas will just skirt around me.

The female, however, has my attention. She is not submerging but is, rather, drifting on the surface, in a behavior the scientists call "logging," and heading more or less in my direction. She has a water bubble on her head. One of the features of orcas' amazing hydrodynamic black-and-white skin, which is part of what enables them to move with such speed and grace underwater, is that water actually flows through it. So when they come up to the surface, especially slowly, you will see them almost wearing the water like a membrane, until they finally burst through and make contact with the air.

This female is doing something like that, moving at a slow and steady pace, almost drifting but actually swimming in a single-minded direction right at the surface, so that the membrane of water remains on her head and upper torso, back to her fin. She isn't breathing through her blowhole because that would break the bubble. It is a dreamlike moment: the whale, in slow motion, heading along at the surface, the afternoon sunlight glinting off the bubble of skin-like water she is playing beneath. It lasts for minutes, a lifetime, and then it is over.

• • •

Orcas live a dream of man. They soar effortlessly, free of gravity, like birds or fairies through the air, gliding above the landscape and observing it from far above. Men have had this dream for as long as they have dreamed. It is why one of their greatest inventions is a machine that lets them fly. It is why, when we create a mythological ideal of a human and call him Superman, one of his chief attributes is that he can fly with grace and ease, as though gravity does not exist for him.

That describes the ethereal daily life of killer whales: gliding sylphlike through their element, their large pectoral fins spread like wings, soaring above the canyons and cliffs of the ocean floor, swooping and diving weightlessly at their leisure, with intelligent minds that rule over all they survey.

Their only real encounters with gravity occur when they play in the element—the air—that gives them such buoyancy in water. Orcas do more than merely breathe air; they leap into it with gusto, in events we call "breaching". It is at once an attempt to defy gravity and to play in it, to celebrate and revel in the earth's pull, to plunge briefly into a world that is both alien and native.

The weightlessness of being is only the most obvious facet of an orca's existence that distinguishes it from what man and his fellow primates experience. Even though they weigh thousands of pounds (tonnage that would render them slow and ponderous as land mammals, as it does elephants and whales' long-distant cousin, the hippopotamus), a whale's daily life is spent in a state of gravity-free seawater, buoyed especially by the lungs that are the legacy of having evolved from land mammals. This state almost certainly played a role in the evolution of whales as the largest creatures ever to roam the planet, since size does not have gravity-related energy costs associ-

A Southern Resident female's exultant breach.

ated with it in a water environment. Whales were free to evolve as large as they liked, even if they sacrificed agility and grace in the bargain, and many species did so, especially the baleen whales, who glide along the surface and with their furry plates eat tiny fish and shrimp floating there.

Orcas are somewhere in between these behemoths and land mammals. Like the larger whales, they are big enough to be extremely powerful, capable of overwhelming nearly everything in the ocean if they choose. However, since they are a smaller whale, they have retained much of the grace, agility, and speed of their fellow dolphins. Orcas can zoom through the water at speeds up to 34 miles an hour and then frolic and play like their bottlenosed cousins.

Besides the relative weightlessness it imparts to air-breathers, one of the important qualities of water, since it is so much denser than air, is that sound travels extremely well in it. Whales in water bathe in sound, and the sounds they themselves make bathe everyone else in it. Loud enough and with enough intensity at a deep frequency, underwater whale songs are known to travel for thousands of miles. Up close and over short distances, water is an even more effective carrier of sound waves—about four and a half times faster than when sound travels through air—and the big-brained dolphin family takes full advantage. It is the medium through which all dolphins not only communicate with one another, but in which they also use sound in a way that lets them see underwater.

Strikingly, orcas (and whales in general) do not hear primarily with external ears like land mammals, though they have small vestigial ear structures on the sides of their heads and highly evolved hearing

mechanisms beneath that. Their chief conduit for hearing is their lower jaw. The lower jaws of cetaceans are not only fairly large, but they are also hollow, filled with a fatty material that collects sound and transmits it through the middle and inner ears and on to the auditory nerve. This structure means that not only are the most distant and subtle sounds detected, but all these sounds are processed by those big brains at lightning speed, which is especially critical for orcas and their fellow dolphins.

That is because they not only are hearing sounds; they are making them at an extremely high rate, and this is not just for communicating but more importantly, as noted previously, for seeing.

Orcas have very good vision with their eyes, located on each side of the head. They are able to see with about the same acuity and for the same distances as humans. However, for any seagoing creature, there is an inherent limit to how far the water itself will allow it to see. Under the clearest conditions, the farthest even a sharp-eyed orca can see underwater is several hundred feet. In the turbid waters of places such as the North Pacific, this visibility can drop to as few as ten feet.

To overcome this, odontocetes—that is, toothed whales— evolved a sixth sense we call echolocation. It is a kind of sonar, although comparing it to the comparatively crude system of sound detection that humans have devised through technology does little justice to how sophisticated a sense it really is. The sounds that orcas and dolphins emit sound to our ears, through a hydrophone, like dense clicks that seem like rapid-fire bullets made out of sound. They make these sounds with a vocal structure inside their blowhole and then emit them in a focused beam through the front of their skulls.

The large melon on the front of orcas' heads is not, as many initially assume, where their brain is located; that is actually farther back in their skull, safely encased behind the eyes. The round melon is actually a large capsule of extremely fine oil, very similar to the sperm whale's coveted oil that came from the so-called spermaceti organ, similarly located above the front jaw. As those gigantic relatives do, orcas use this oil sac for focusing and transmitting those sound bullets into the water. The melon is actually a lens for seeing with sound.

The sound bullets that come out of the melon are very dense packets of sound, not entirely dissimilar from the packets that transmit computer information electronically, and so when these sounds bounce back to the killer whales, they carry a broad spectrum of information. This apparently means that when this information is processed by that highly evolved and complicated brain, it renders the orcas capable of not merely detecting the presence of objects (as our sonar does) but rendering a clear and detailed vision of what is there in the water. Indeed, it goes beyond mere vision; orcas can see *inside* things. Because sound is actually capable of penetrating objects better than light, the bounceback from orca echolocation includes the more subtle variations that occur as the sound penetrates an object and then returns.

Experiments conducted with captive orcas, for instance, have established that they (and other dolphins) are able to detect the nature and shape of objects hidden inside opaque containers. There are a number of anecdotes from female dolphin and orca trainers whose subjects began acting differently (in one case, protectively) around them, leading the trainer to later discover that she was pregnant. These latter incidents, however, are purely anecdotal, as with so many things dolphin-related; there has never been a proper scientific study to determine if dolphins can detect human pregnancy.

"I think it's extremely plausible that dolphins, including orcas, would be able to detect a fetus," says Lori Marino. "We know from other studies that they are very good at going from a visual image to an acoustic image," and vice versa.

The sophistication of dolphin-family echolocation is part of why the United States Navy employs dolphins to locate objects that can't be found otherwise. In the summer of 2013, for instance, Navy dolphins were able to lead divers to an antique 19th-century torpedo that was then recovered from the ocean floor off the coast of California, not far from the Hotel del Coronado in San Diego. It was a find, Navy officials explained, that they could never have made with their own multibillion-dollar sonar systems.

"Dolphins naturally possess the most sophisticated sonar known to man," boasted Braden Duryee of the Space and Naval Warfare Systems Center Pacific. The dolphins who found the torpedo were being trained in mine detection, and when they kept alerting their handlers to the presence of an object where none was supposed to be, they were asked to place a marker on the spot. When divers investigated, they found the old torpedo.

Orcas' sonic capacities have also inspired human efforts to replicate them. Researchers at Stanford have developed an ultrasensitive hydrophone modeled after the design of killer whales' middle ear, which employs a tympanic bone-and-plate complex to transmit sounds into the nervous system. The device the scientists constructed based on that model now allows researchers to capture a wide range of ocean sounds, from up to 160 decibels to the quietest whispers, and to do so accurately at any depth, something that was never possible with traditional hydrophone designs.

"Orcas had millions of years to optimize their sonar, and it shows," explained Onur Killic, one of the researchers. "They can sense sounds over a tremendous range of frequencies, and that was what we wanted to do."

Naturally, orcas are sophisticated not only at making sounds but at listening to them, too. As with everything killer whales do, it is shared, a social thing.

• • •

The female orca is not only drifting intently toward me with a water bubble on her head. She is also echolocating me. Like crazy.

The sounds that orcas emit when echolocating are striking. We call them clicks, but that's not quite right. Unlike the calls and vocalizations we all know from various whale-sound recordings, these sound bullets go rat-a-tat-tat-tat-tat through the water, and you know when the bullets are being directed at you, because they become intense and direct; you can practically feel them. At times, the bullets fly a second or so apart, and then, when an object gets their attention, they will let off whole strings of them: br-r-r-r-r-r-r-rt.

The female watching the calf is pounding my kayak with these sound bullets, but she is not the only one. In general, these orcas I am watching at Kaikash Creek are being very businesslike under the water. Listening on my hydrophone, I am struck by how few vocalizations they are making. Instead, the clicks and chains of clicks are everywhere. It would be reasonable to hypothesize that what I am hearing is the whales busily hunting, rather than socializing. Since we can't really see what they are doing underwater, however, we can only guess. What is fairly obvious is that this female is echolocating me heavily because she wants to keep an eye on me. Or perhaps more importantly, she is trying to alert the calf to my presence, as well as my position, so that it can avoid coming into contact with me.

Then it strikes me. Echolocation is something that whales not only use for themselves, but something that they can share. The calf gets bounceback from the same sounds that its mother is emitting. It is almost as though she is shining a sound flashlight on me so that her young charge, preoccupied with chasing fish, is aware of my presence and doesn't get too close.

Tests with captive dolphins, have found the same thing. Dolphins were found to have the ability to "eavesdrop" on another dolphin's inspection of an object. This may also help explain their social behavior when hunting, since the more orcas or dolphins there are lighting up parts of the water with their echolocation sounds, the better they all can detect and snag their prey.

It suggests, moreover, that echolocation is more than a mere sense. In the hands of orcas, as it were, the sense becomes a social activity, even a kind of communication.

Males can intimidate with just their massive dorsal fins.

Eventually, the calf rejoins its overseer a good twenty yards away from my kayak. I am so focused on these two that I have not observed the large male coming up from behind. All six-foot black fin and spray, he bursts up near the female and her calf. My heart jumps. If there's one thing these creatures are good at, it's surprising people.

The other striking thing about these Northern Resident whales is how little vocalizing is going on. This is somewhat unusual in my experience, which is mostly with the rather gregarious Southern Residents. On this day with this pod, there are very few of the squeaks, squonks, cries, calls, and various odd noises that killer whales produce in the process of communicating with each other. These whales are all business, almost exclusively echolocating, seeking out the same Chinook salmon that are similarly the favored prey of their Southern counterparts.

I had often heard about Northern Residents that they were less playful than Southern Residents, less prone to the demonstrative behaviors such as spyhopping, tail-lobbing, and breaching that were the trademarks of their Southern relatives. I turned my attention briefly away from the big male and focused my camera on the female and calf, who were still relatively close to my kayak, musing to myself: It must be true—they certainly seem less demonstrative.

At that moment, no more than thirty feet away, the big male leapt fully out of the water in a high breach, looked at me, and came back down to the water in a massive splash. I gaped. The splash was the only thing my camera, far too late and too slow into action, captured.

Surprises. Orcas have a knack for them.

• • •

Even the way that orcas sleep is social, something shared.

I was still sipping my morning coffee one of the mornings at Kaikash Creek when a pod of orcas, sleeping, came past. You could tell they were asleep because the procession—*kooosh kooosh kooosh* in succession as they broke the surface of the glasslike water, rhythmic and slow and majestic—was clustered together closely, in a line, and surfacing only every forty-five seconds or so, coming up and going down as a group. I sat on my log and watched them. The only other sounds were the keening of the gulls that occupied much of the rocky beach.

Every breath that killer whales take is voluntary and conscious; unlike most land mammals, most cetaceans do not have an involuntary-breathing mode. Early human captors discovered that when they anesthetized dolphins and orcas in their care, it killed them, because they simply stopped breathing. Sleep, however, is a physiological requirement of every mammal on earth. So, when and how do they manage it?

In the case of killer whales (and most dolphins), the trick involves shutting down only half of their brain during a given sleeping session, remaining just awake enough to swim, surface, and breathe, all in a slow, rhythmic pattern. In the wild, where orcas swim almost constantly, this is done in large pods of up to twelve whales, who line up in a wide

A closely clustered group of traveling Southern Residents.

arc and draft off each other's wake as they rise, breathe, and submerge together.

This is something spectacular to see, even if the orcas are not at all playful when sleeping. There is nothing quite as moving as the sight and sound of a long row of killer whales surfacing one after another, the plumes hanging in the air, and then vanishing below the surface when the procession finally reaches the end of the row. The silence follows, hanging there, for somewhere between 15 and 20 seconds, and then as certain as a clock, the first in line resurfaces again, a little further along on the same line of travel, and the whole row rises behind them, like fingers rising and falling on an oceanic piano, playing a syncopated melody.

The one constant in orcas' lives is their togetherness; sleeping and awake, eating and playing, traveling and exploring, everything is done together. Every ecotype of killer whale has been observed engaging in a form of prey sharing. Resident killer whales share salmon they have caught with other whales; transients and Antarctic whales share seal kills; North Atlantic orcas team up to herd herring into balls; and New Zealand whales who hunt rays on the seafloor often team up to hunt them and then share their meals with one another. Moreover, these are animals who mostly remain with their familial group for every day and moment of their lives.

Socializing is wired into orcas evolutionarily, at a level that dwarfs the comparatively loose social ties of humans. One of the logical consequences of this is that it requires an abundance of the quality that makes social life possible: empathy. By empathy, we mean not simply the ability to sense what other people and beings may be feeling, but to feel it ourselves, and then to act accordingly, perhaps contrary to one's immediate self-interest or desires. It is empathy that not only makes orcas most like humans, but perhaps makes them more than human. For modern humans, empathy is not a universally desirable trait, since it reeks of vulnerability in an ever-competitive world. For killer whales, empathy is an evolutionary advantage.

• • •

These same whales hang about these waters for the next several days, evidently drawn by the chance to rub themselves on the smooth pebbles at Robson Bight, an area specially protected by the Canadian government for the orcas' use. Every summer the unique qualities of these rocks draw Northern Residents to these waters, where they can be observed rubbing their tummies and bodies on the time-worn rocks that comprise the beach at the Bight. No one is quite certain why they do this, but it appears from observing them that they enjoy the massaging effect of the stones.

It wasn't as if these whales wouldn't vocalize. Two days after my initial encounter with this pod, I spent over an hour listening to them talk to each other. The strange and mysterious part of this was that I really couldn't tell where they were at the time. It was late in the afternoon, and I was hearing on my radio from several whale-watching operations that several pods were about to converge in the waters near Kaikash Creek. So I hopped in the kayak to see what I could see and hear what I could hear. Sure enough, just a little offshore, I picked up orca vocalizations through my hydrophone.

They were distant at first, and I certainly could not see any orcas on my immediate horizon. One of the reports indicated that whales might be approaching from the north, and the current that afternoon was headed strongly northward. So, I just drifted with it to see what came along. All this time, I kept hearing orca vocalizations, distant, but increasingly clear. I thought perhaps they were coming closer, but no black fins ever showed themselves on my northern horizon.

I also knew that the pod of whales I had encountered was supposed to be lurking in the area of Blackney Pass, a passage between Hanson and West Cracroft islands directly across Johnstone Strait (about two miles) from my camp. I peered over that direction with my binoculars. It seemed I caught a glimpse or two of a black fin, but it was hard to be certain, since they were obscured by the reflective mirage that the water was throwing up along the distant shores of the islands. I looked southward, just in case—orcas do love to surprise us—but there was nothing there, either. Strangely, the sounds became increasingly clearer. Perhaps this was because there were very few motorboats out in the Strait or perhaps it was because the whales began vocalizing more and more loudly.

Whatever the cause, I wound up with nearly two hours' worth of recorded orca sounds from that afternoon. It was lovely on the water, although after a time I had to begin paddling back southward before the current carried me all the way to Telegraph Cove. The whole time, I was listening to a cacophonic chorus of orcas chattering away, or whatever it was they were doing—singing, perhaps? Or perhaps something more mundane, such as talking about the currents? Or, more likely, communicating in a conceptual way that we cannot comprehend.

Even though I could hear them with great clarity, there was always a distant, echoing quality to the sounds that told me they were coming across at some distance from Blackney Pass, like songs heard from across the deep canyon that lay under the waters of Johnstone Strait. That same quality gave the calls a strangely haunted beauty, as though they were coming out of the past or some other way through time.

The whales never moved for all that time, remaining elusively out of view. So, I never did see the whales that day. After more than two hours, I paddled back to camp and made dinner, the haunted cries still ringing in my ears.

• • •

Whatever is going on among orcas and their vocalizations (and we don't really have a clear understanding of this at all) the one thing that is self-evident is that this is their primary means of communicating with each other. *What* they are communicating and *how* remain grand mysteries. It is probably the most tantalizing aspect of what we know, and don't know, about killer whales, because if orcas are actually conceptualizing abstract ideas at a high level, and perhaps even retaining long-term intergenerational memories, if these communications really are a language, then the possibility of interspecies communication becomes real, and all the speculation about their intelligence is substantiated. However, we simply don't know whether this is the case or not, and the evidence to date suggests that the barrier may well be unbreachable.

The immediate problem with orca communication is its apparent limitations, at least from our human perspective. The stereotyped calls with which they communicate are fairly complex by animal standards but have serious limitations, if judged on the basis of human communications, since they appear to have an extremely small vocabulary.

There is a constellation of about ten traits that define what human linguists mean by language, notably limitless expression, a discrete combinatorial system (using a system of symbols whose coagulation forms disparate meanings), recursion (the internal embedding of syntax), and a memory system, as well as the creation of new sounds, arbitrariness, and the ability to convey information based on perceiving in advance what the listener already knows (called "social cognitive aptitude"). As Justin Gregg has persuasively demonstrated, bottlenose dolphins score reasonably well on many of these counts (cumulatively, they score a 20, about the same as chimpanzees, on a scale in which humans form a baseline top score of 50), but score a zero on the definitive issue of limitless expression.

"When you look for the markers of what we call language," says orca scientist John Ford, who studies their communications, "they simply are not present."

If you want to know why no one has ever deciphered dolphin communications so that we can talk to them, and they to us, the answer is simple: The scientific community reached the conclusion some time back that these communications did not constitute language, and there has been very little evidence to emerge since to alter that conclusion. It was simply not considered a fruitful avenue of investigation.

However, whether this is the case for killer whales is an open question, since they have been studied very little in laboratory settings; most studies of orca calls have been in the wild, and most of them have focused on social aspects of the communications without delving into structure and meaning, since it is so difficult to observe their underwater behavior, especially in relation to their communications. If orcas are only slightly larger versions of bottlenose dolphins, as they are frequently viewed, then their communications are also of limited scientific interest. If, instead, they are to dolphins what humans are to chimpanzees, then it's possible that science has so far overlooked this avenue.

It is important to note that, in addition to the differences in their physical brains (orca brains are not only four times larger but a good deal more complex than bottlenose brains) scientists have observed an important qualitative difference between dolphin and orca communications. Orcas communicate in a socially structured way. Whereas dolphins collectively seem to chatter willy nilly with little regard to one another, orcas both captive and wild are generally more polite. They tend to wait for each other to finish calling before making their own calls.

Orca call exchanges thus take on at least the surface appearance of conversations, even if only crude informational exchanges. Many of these exchanges feature idiosyncratic sounds, the meanings of which are utterly unknown. This has implications, moreover, for the linguistic possibilities of orca communications, since the existence of conversation also suggests the possibility of exchanging ideas.

That is hardly the end of the story. Orca calls are in fact very complex and acoustically rich, suggesting that information could be imbedded within them in ways that are imperceptible to human ears, just as the peculiar sounds of a computer modem, heard over a phone, sound like mere repetitive screeches but are, in fact, carrying signals rich with information. You just need the right processor to translate it all. Theoretically, this could be what is happening with orca calls, and we would need an orca's complex brain to be able to process and translate it.

This remains, however, largely a matter of speculation. The truth is that, while dolphin communications have been studied extensively, work on killer-whale communication has been limited in scope and depth. If killer whales' communications are not significantly more sophisticated than those of their cousins, the bottlenose dolphins, then the picture is ambiguous. Dolphins have shown a clear ability to learn a simple human language and to comprehend syntax, but their poor scores when tested for infinite expression make it seem unlikely that they use what we call a language.

Orca vocalizations typically are a fairly limited set of calls, usually comprised of a single rising and falling tone that is specific to each orca's familial pod, with similar tones appearing in sequence, repeated between individuals in an apparent call-and-response behavior. These are accompanied by idiosyncratic whistles, grunts, and squawks. The meanings of each of these sounds is utterly unknown, since no one has ever been able to track wild orcas' underwater behaviors in relation to the sounds they are making. Even among captive orcas, there has been nothing to indicate that these are in any way different than the calls emitted by birds and other mammals, other than that the sounds are specific to family and broader social groups rather than to entire species.

All told, the scientists who collect these sounds typically have found about forty different sounds that wild orcas will make in their social interactions with other orcas, and these sounds are usually specific to a single family group. This appears to be a fairly limited vocabulary, at least on first hearing. Any linguist hoping to catch glimpses of language therein may well be disappointed. Nonetheless, the distinctive social-group-oriented nature of the calls is deep enough that even scientists are comfortable referring to them as "dialects."

A simple interpretation that is clear to the scientists who observe them regularly is that there is almost certainly a beacon-like quality to the calls; their frequency during travel and foraging suggests that these calls let the orcas all know the positions of their fellow pod members and perhaps enable them to coordinate hunting behaviors. Another purpose may be simply reinforcing pod identities, since the call repertoires remain stable over generations within these pods.

The most seasoned analyst of orca communications is Canadian whale scientist John Ford, the man whose research established definitively that not only do orcas use calls for communication, but that they have dialects that reflect their sophisticated social organizations. Ford has spent years analyzing orca calls and echolocation in the wild, and he remains, after all these years, a little baffled and awestruck—but realistic, too.

"In my opinion, the communication is a challenge for us," says Ford. "There's much more to learn about it. We don't have all the answers. But I've yet to see any evidence of representational signals— the features of what we call language. I think it's pretty much a here and now communication system."

Typical of scientists, he qualifies these judgments by noting that linguistic features in orca communication may in fact exist, but there's no data to support it yet, because "so much communication goes on with these whales under the surface where we can't see what they're actually doing, it's hard to link the fine details of behavior to the calls."

"In my opinion, the different repertoires of the calls that are used by these groups are most importantly kinship flags, or badges—they convey really sophisticated information on the genealogy of the individual and his group, and play an important role in keeping the group together, but also in keeping them from becoming inbred, by basically encouraging cross-breeding outside the dialect group," Ford says.

Ford, too, has observed the echolocation-sharing qualities of social orca pods: "No doubt they do share echoes," he says. "In fact, that's another study that Lance Barrett-Lennard did, looking at echolocation differences between residents and transients, where he was able to plot the amount of echolocation activity against increasing group sizes in residents. It's not a linear relationship at all.

"So the simplest explanation for that is that the proportion of echolocation activity decreases with increasing group size, so it would suggest that not everybody is swimming around clicking independently—they're making use of each other's clicks. Or they're just following the animals that are doing the clicking.

"They certainly can get information from their mother's echolocation, as you experienced. Because the clicks are so directional, though, it's only going to work when the animal eavesdropping is really close to the one clicking. I think there are a lot of social components to echolocation as well."

There well may be more to orcas' vocalizations than the simple communication we hear. Scientists who have analyzed the nature of the calls say that, as with the echolocation bullets, these are very dense and rich sounds. Even within the range of orcas who are repeating their stereotyped family calls, there are tremendous variations in tones, in intensity, and in volume. There are also variations in emotional content. It is not unusual, for example, to hear an adult seeming to chastise a young calf it is accompanying with a series of demanding-sounding chirps (though whether that is the actual intent is hard to say). Repeated stereotyped calls also vary widely in intensity, at times sounding more urgent and emphatic than the repeated beacon calls that populate many of their conversations. Considering that we are talking about animals whose sense of hearing is far more acute and wide-ranging than ours, and whose brains are clearly capable of processing extremely high levels of data from that sound, it is not unreasonable to suspect that there is much more to orca calls than meets the human ear.

Val Veirs, a retired Colorado College physicist who now monitors Southern Resident orca calls from his home on the western side of San Juan Island—as well as all the sound events in Haro Strait—with a sophisticated array of hydrophones and computers, has developed his own theories about what might be going on with killer-whale communications. He has paid especially close attention to the well-known stereotypical call of the J2 pod, known as S1.

"My feeling is that S1 is like a radio station for the J pod," says Veirs. "When we dial up 91.5 here, we get the Port Angeles NPR station. And at 91.5, that's like S1; the call is like a channel. It's a channel that J pod has decided to use to communicate. Now, when you listen to 91.5, you're not listening to 91.5 million cycles per second—that's pretty irrelevant to us. It's important to the radio, but there's no information for us encoded in that, the information's encoded in the way the frequency shifts a tiny little bit, and you'll get FM radio.

"So I think something like that may be going on with orca calls. When you look at S1, the sound is not exactly the same every time you look at it. On the first order, it doesn't sound the same. But if you look at them on a spectrogram, some of them are extended, or there's wavering in the harmonic; some of them have more in the fourth or fifth harmonic than they do in other calls."

Orcas also have demonstrated a capacity for mimicry. Scientists recording the sounds Northern Residents make began picking up the odd appearance of calls from other pods among whales that would socialize with those pods and not only when those other pods were present. At odd times well away from those pods, orcas were heard making their stereotyped calls to one another.

"Mimicking another group's calls could be a way of referring to that group . . . or of communicating something about that group to one's own family members," said behavioral biologist Brigitte Weiss, who conducted the study, noting that it's just as possible that these calls have no function at all. "Most likely, the answer lies somewhere in between."

• • •

Perhaps the most striking case of orcas' sound mimicry capacities involved a little whale named Luna. It was also one of the most striking manifestations of orcas' empathy. Luna was a two-year-old male from the L pod of the Southern Residents—his number was L98—who showed up alone, quite mysteriously, in Nootka Sound, on the western side of Vancouver Island, in the summer of 2001. His appearance there coincided with the sudden and worrisome loss of four of L pod's young males in the course of a single year and may have been connected to it, since one of the young males who died was Luna's frequent companion. The suspicion is that sometime that winter while the L pod was foraging along the Continental Shelf, when the male was dying, Luna and the male wandered away from the rest of the pod, and then L98 found himself alone, after his companion died, in unfamiliar but protected waters with lots of available food.

The calf soon began making his presence known among the people who lived in Nootka Sound, notably those who used boats: fishermen, leisure boaters, loggers, and water-taxi operators. The affable and playful young orca would approach people in these boats and then, when they would turn off their engines, come right up to the edge of the boats and let people stroke and pet him. He would play around the boats, much to the delight of most of the humans who encountered him.

Luna in particular befriended the crew of a boat, dubbed the *Uchuck III*, a lovingly restored old wooden steamer painted sleek glossy black, that ran passengers and supplies up and down the Sound. At first, the boat slowed down when Luna approached, and its crew became familiar faces for the whale. After a while, the *Uchuck* would continue on its run, and Luna would follow, playing in the boat's wake and delighting both crew and passengers. He also came up to local docks and engaged with people there. He was fond of playing with hoses that were lowered into the water within his reach; he would grab the hoses and squirt the water around, often hitting his human friends of the moment, who would laugh uproariously.

Not everyone was quite so enamored, however, particularly not the loggers and fishermen who were trying to work when Luna would interrupt them. Pretty soon he became a problem for the Canadian government's wildlife biologists, who were concerned that his frequent contacts with humans would make it more difficult to reunite him with his family, as was their then-vague hope. They were also concerned that his frequent contact with boats might lead to Luna's being injured by a bow or propeller. Complicating matters was that Luna had become a tourist draw. People began showing up at Gold River, the main town reachable by road on Nootka Sound, hoping to get a look at the whale, perhaps to even interact with it, and locals began taking them out in their boats so they could do just that.

The Department of Fisheries and Oceans (DFO) announced it was instituting a policy of forced separation, requiring humans to keep their distance from the little whale and not to interact with him if he approached their boats. At first, this was enforced by the DFO's over-worked officers in Nootka Sound, but soon the job of enforcing the rules was placed under the control of the Luna Stewardship Project, led by a government biologist who had experience with orcas. This project employed private citizens to act as whale stewards. They would go out in pairs—most of the time, it was two women—in an official-looking boat and try to remind boaters not to interact with Luna when either they or he approached the other.

This turned out to be much easier said than done. Luna was ac-tually very aggressive about seeking out human contact, and some of his favorite contact, as time went on, was with these stewards. If the DFO officers approached boats that were getting close to Luna, he would actually get in between them and force contact with both sets of humans. In one of the videos capturing these stewards in action, Luna can be seen spyhopping in between the two close craft and nuz-zling the hapless steward, who is holding onto the side of the approach-ing boat from her own, even as she chastises its owners for getting too close and coming into contact with the whale. She can't help but laugh, "Luna, you're not helping!"

Eventually these stewards changed tactics, deliberately cultivat-ing a relationship with Luna so that they could draw him away from other boaters and into unpopulated parts of the Sound. However, then the DFO officials in Victoria switched back and ordered the stewards to return to enforcing the rules they had created keeping boats separate from the whale—with, once again, a predictable lack of success.

Luna not only attracted tourists; scientists, too, were drawn to him, because the opportunity to observe a wild orca so closely provided unique data-collection opportunities. It was during one of these studies that he was observed mimicking not other orcas, but sea lions.

A team of researchers led by Andrew Foote recorded a number of

California sea lion sounds while out in Nootka Sound following Luna. However, they also recorded a number of sounds that *sounded* like barks to the naked ear, but which when analyzed were not in the same harmonic range used by sea lions, but were instead in the range used by killer whales. Some of these barks were recorded when there were no sea lions observable, and only Luna was present. It was apparent that he was imitating their barks, perhaps calling to them, as though they too kept the lonely young whale company.

That was the adjective nearly everyone who encountered Luna used to describe him: lonely. He craved company, and the humans were happy to give it to him, DFO rules notwithstanding. Many of the locals took to looking after the young whale and petting him at the docks. One of them was ticketed by the local police for doing that and went before a local judge about the matter. She wound up with a $100 ticket.

Not that she regretted it. "Not at all. It's the best $100 I ever spent," she said, smiling.

"When you see wild whales, they look at you, but they carry on," observed another longtime local. "This is a whole different thing. Because he doesn't just sort of go on by and do his business; he comes up and looks at you and shows you stuff. He's flashing his tail and doing all this within fifteen feet of your boat. He is trying to communicate with you."

The sea lion barks were not the only thing that Luna mimicked. Two journalists who arrived in Nootka Sound in 2002, Michael Parfit and Suzanne Chisholm, found they had the same difficulties avoiding contact with Luna as other boaters. He was particularly fond of pushing Parfit's little inflatable boat around and sometimes played with it at the dock.

One day Parfit filmed Luna playing with a small aluminum outboard boat at the dock, placing his rostrum next to the hull at the boat's rear, near the little motor, and imitating it loudly, using his blowhole like a pair of lips: "Brrrrrr-r-r-r-rrrr!" Funny thing: it really did sound just like an outboard boat engine.

There was a brief moment in 2003 when an opportunity arose to reunite Luna with his home pod. Ken Balcomb of the Center for Whale Research showed up with a boat that was capable of making the long journey around to the San Juan Islands, equipped with sound-making gear that could entice Luna by playing L pod calls, with a plan to get him back to the Salish Sea and his home pod. However, the plan entailed more human contact with Luna, and so the chief bureaucrats at the DFO vetoed the idea. Balcomb was warned he would be arrested if he attempted any contact with Luna, and so the plan was shelved.

In the meantime, the First Nations people of Nootka Sound—the Mowachaht/Muchalaht tribe—became deeply involved. The year before Luna had appeared, a tribal chief named Albert Maquinna had passed away, and Maquinna had told friends and relatives beforehand that he planned to return to the tribe as a *kakawin*, a killer whale. Orcas are not particularly common inside the Sound, so Luna's appearance was taken as a certain sign that the chief had lived up to his word. The tribe by treaty right had some say in the matter. They approved of the plan to reunite Luna with his home pod and made clear that they would fight to prevent the whale from ever being placed in captivity.

Public pressure was mounting and after years of trying to simply manage Luna, the DFO in the fall of 2003 announced plans to undertake a major operation to reunite Luna with his home pod, with funding help from American sources. The plan was to capture him in an enclosure, lift him onto a truck in a sling, and drive him to the San Juan Islands so that he could be placed in a sea pen there and eventually placed with his L pod family members. After years of foot-dragging, the action was welcome, but then DFO announced that it would wait until the next summer to actually enact the plan, overruling the advice of their own scientists to repatriate Luna as soon as possible. Then word leaked out that DFO also was negotiating with marine-park officials to place him in captivity if the reunion with L pod didn't take, a situation laden with incentives for the captors to declare Luna unable to return to his family, since captive killer whales are worth millions of dollars. On top of that, DFO began shutting out the various interests involved, including the Mowachaht tribesmen with whom they simply refused to negotiate. The respected researchers at Paul Spong's OrcaLab were likewise told they would not be permitted to participate, even though they were the resident orca experts on Vancouver Island and had worked with DFO for years, including a previous successful reunion of a lost Northern Resident calf with its mother. Suspicion festered like a bad algae bloom. As the day planned for the capture approached, matters worsened. There was a report that officials at the Vancouver Aquarium, who were overseeing the capture, were warning local medical authorities to prepare for tear-gas injuries. They had closed off the airspace over the sea pen they were setting up near Gold River to prevent media from reporting too closely on the capture.

So on June 14, 2005, the day that the DFO had designated as the day to draw Luna into their sea pen, a crowd gathered onshore in Gold River alongside the TV news crews and observing biologists, and more importantly, a flotilla of traditional tribal cedar canoes gathered in the water outside the sea pen. It was a sort of protest, but all the tribal members did was sing songs to the whale, bidding it farewell. However, it proved to be a fateful distraction. As expected, the captors were able to persuade Luna to visit the pen, but only for brief periods. He was far more interested in the people in the canoes singing to him, and so he quickly exited the pen and went out to join them and play with them. Occasionally he would wander back to the pen, but never long enough for them to enclose him. After hours of the cat-and-mouse game between the tribal protesters and the would-be captors, they closed down.

So it went for eight more days. Luna would visit the sea pen, but he was never deeply interested in it. The singing humans, who showed up every day in their canoes, held much more attraction for him. At one point, the captors appeared to have Luna secure inside the pen, but he dodged under the small boat they were using to herd him at the last second, then surfaced on its other side, pushed the boat into the pen, and then swam away. Finally the DFO gave up, acknowledging they had been outmaneuvered by the tribal members whom they had so needlessly disrespected. They then reverted to their old policy of simply trying to manage Luna's behavior and hoping his family would come past Nootka Sound. A tribal biologist took up the job of providing Luna with human company away from other boats, and the orca's life settled into its old pattern of disquieting solitude intermingled with human interactions.

In all, Luna spent five years in Nootka Sound. He was clearly well fed throughout, meaning that when it came to finding food whatever he had lost as a pod member, he was able to compensate by hunting opportunistically (there was evidence he ate all kinds of salmon, unlike his family, which is known to hunt Chinook salmon almost exclusively). During that time, it became clear that for his social needs he had replaced his orca family with the variety of humans he encountered. In the process, he opened a window into the world of killer whales for all the humans who came into contact with him to peer through. For those who were fortunate enough to get that glimpse, it was life-changing.

Michael Parfit and Suzanne Chisholm, who were both married and professional journalistic partners, had come to Nootka Sound on a short-term assignment of a few weeks for *Smithsonian* magazine and wound up staying there for three years and making a documentary film about Luna. They, like so many other people in the Sound who had encounters with Luna (including the DFO officer who had originally been in charge of him), became emotional captives of the whale almost from the very first day that they met him.

It happened early in their stay, during the period when the DFO was working hard to keep people separate from Luna. They had gone out to see if they could find him in their little boat when he exploded out of the water beside them. Luna loved to play pranks like this, loved to surprise his human family. However, these humans hurried to shore, worried that they were not supposed to interact with the whale. He followed them there and lingered in the water next to their boat, peering up at them.

"There was such consciousness there, such deliberate intent to connect; it was just this unforgettable moment," said Parfit later.

Eventually, the couple came to champion Luna's need to have social contact with humans, shedding the mantle of journalistic objectivity, because it seemed the right thing to do. They were aware of orcas' large brain capacities, and their personal experience with this one convinced them that "maybe he had empathy."

This realization rubs against old-fashioned scientific approaches that insist on never projecting human traits onto animal subjects—anthropomorphizing them—but Parfit believed that empathy was only logical, considering the highly social nature of orcas in the wild. "That seemed like a very human thing to expect in an animal," he later wrote in his deeply felt account of the Luna saga, *The Lost Whale*. "But not really. Scientific studies have shown that something like empathy is at work in several species. And when I thought about the social nature of orcas, it turned out to be obvious. If an orca has social needs, he also has to have something that at least resembles empathy."

It was, tragically, also Luna's eventual undoing. On March 15, 2006, he was playing with a tugboat he knew well, an ocean tug named the *General Jackson*, when he got pulled into the propeller and was killed almost instantly. His body sank to the bottom of Nootka Sound. Luna's death made headlines throughout the Northwest, and there was an outpouring of grief from whale advocates around the world. Three months later, his human friends in Nootka Sound put together a memorial for the whale. Flowers were cast on the waters of the Sound, and tears joined them in a last, tender farewell.

Michael Parfit summed up the thoughts of many of the people who had been touched by the whale. "What the experience with Luna gave us is a sense of the fellowship of the wild that we're part of—that sense that we're in this together with these beings who are out there longing and suffering and expecting joy and going through something that is not so different from us, and yet is, in amazingly mysterious ways that we can only approach in an empathetic way, but not a way we can actually pin down," he told a radio interviewer.

"We're looking at these [animals'] lives and we describe them as being inferior to us, when in fact their lives are so rich and complex and their awareness of living is so present," he added. "In a very small sense, what Luna was saying is [people] are not paying attention to what he was asking for. We need to learn how to respect these other members of the fellowship and try to learn what they need."

• • •

One June morning I went for a paddle on the west side of San Juan Island, again not expecting to encounter any whales. It was early, before many boats were out, and it was a classic paddle; the water was quiet and glasslike, and there was a grey fog shrouding the water and the shoreline. Gnarly, red-amber madronas spread their strawberry limbs into the air along the rocky shore, like ancient sentinels reaching up to pluck a cluster of green from the mist. The only sound was the gentle lapping of the waves on the granite cliff wall, the slurping in the kelp beds.

I spotted the pod of whales approaching from a distance. The initial group that went by me was about a half-mile out in the channel, and they appeared to be in transit mode, chugging along rhythmically at about 15 knots, rising and falling in a predictably rapid pace. Nonetheless, I tucked into a cove along the rocky shore, amid a cluster of bull kelp, throwing one of the fronds across the deck of the kayak as an anchor, and rolled out my hydrophone, hoping I could listen in as they passed by. I had learned long before that the sheer cliff that drops away below the surface of the water along this stretch of the island makes for some interesting acoustics.

The first group wasn't making much in the way of sounds, and I dropped my headphones around my neck, waiting to see what else might be out there. Bald eagles are plentiful along here, as are harbor seals. Occasionally you will see river otters and minks as well, and there were probably going to be more whales. You will almost never see a lone orca in these waters, and even threesomes are unusually small groups. These are very social animals. Their home is wherever their fellow whales are.

"*Kooosh!*" Once again, I heard the whales before I saw them, but this time they were close; a group of four whales, in next to shore, hunting, was coming just around the bend from where I was tucked away. There were two males, a female, and a calf.

The female, I would ascertain later, was a 17-year-old J-pod orca named Polaris (her official number is J-28). She's readily identifiable by the notch (probably from a boat propeller) in her dorsal fin, and the calf was her first baby, a female numbered J46. At that time, the calf was only a year old, but that also meant the baby was now old enough to be given a name. Mortality rates are high among resident orca calves, for a number of reasons, so the scientists who study them don't give

The calf J46, aka Star (left), nuzzles with her mother, J28, aka Polaris.

them names until they survive their first year. They had named this little one Star.

It was an apt name. Like many calves, she was playful and curious. She kept heading in the direction of my kayak, and her mother kept intervening, distracting her to other, more appropriate interests, such as salmon.

They appeared, in fact, to be hunting; Polaris was lingering near the surface, seemingly intent on her prey, and then she made a quick turn and disappeared beneath the surface. Star followed, and all was quiet for a moment, except for the two large males who lurked on the perimeter, like playground guards patrolling. Then Polaris and her baby burst together out of the water, rolling and rubbing. The mother appeared to have a salmon in her mouth, and as they rubbed closely in the water, she appeared to be sharing it with her calf. Then they both submerged again.

When they reappeared forty seconds or so later, it was to resume the close-contact play in which they were indulging. Star was rubbing all over her mama, and the mama in turn was nudging the calf about playfully. At one point the mother nudged her rostrum out of the water in an extended spyhop, while the little one rose up alongside her, half out of the water, and peered about, creating the brief illusion that she was every bit as big as her mother.

It was a tremendous and touching display of affection. As I watched them, I recalled my then-fading experiences as a first-time father with a new baby, now grown; how she had craved the physical contact, how she had loved to crawl all over me, and how I had loved it in return. However, that kind of bonding occurs at a visceral level, and it's universal to humans and animals alike. Polaris and Star were being profoundly human and profoundly wild as well. For me, at least in that moment, experiencing that kind of deep common ground with a creature in the wild did not make them more human. It reminded me, instead, that we are all animals, and that is not always a bad thing. Love and affection, loyalty and kindness, perseverance and mercy: these are things we know in animals, too, after all.

Familial, or at least pod, loyalty was at work that morning, too. The two males who appeared to be guarding this twosome kept lurking in the vicinity. One of them, a big 19-year-old named Mike, numbered J26, with a fully sprouted, five-foot dorsal fin, came swimming closely past my kayak, wrapping his dorsal in the kelp fronds near my boat and then swimming on, as if to remind me that I really wanted to stay tucked where I was. I took the suggestion.

Mike's brother Keet, who was born in 1996, was the other male hanging off the shore. These whales were not part of Polaris's direct family (Princess Angeline's J-17 matriline), but they were part of a subpod that historically accompanied those orcas, namely, that was headed by the 39-year-old matriarch, Slick, aka J16. They probably were hunting, too; the run of Chinook salmon that spring was strong and plentiful. Keet seemed more shy and wary or perhaps just more intent on the day's meal.

After a while, the group moved on northward, following at their own pace the fast-moving pod that had led them this way. Then they passed around the next bend in the cliffside waterline, and I could hear them no longer.

I tucked away my hydrophone and my camera and pulled out into the back-eddying current that was now beginning to swirl. It pulled me back into camp a little while later, where my wife and daughter were waiting with coffee and breakfast and welcome smiles. Fiona is too big to squirm all over my chest, head, and shoulders, as she once did, but she still gave me a warm hug that morning, and I could not help but revel in it, like Polaris and Star. Some things are universal.

North Pacific Resident ecotypes

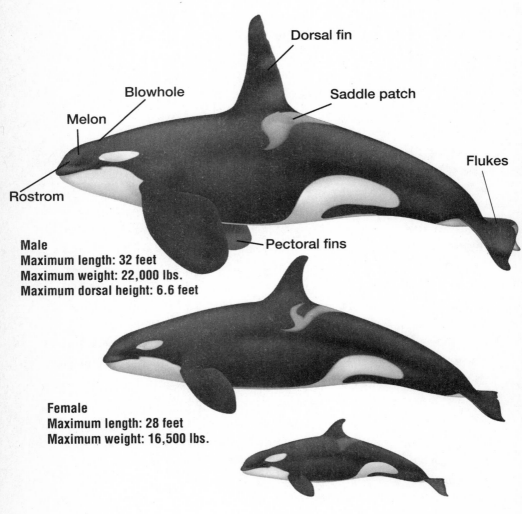

Dorsal fin

Saddle patch

Blowhole

Melon

Flukes

Rostrom

Male
Maximum length: 32 feet
Maximum weight: 22,000 lbs.
Maximum dorsal height: 6.6 feet

Pectoral fins

Female
Maximum length: 28 feet
Maximum weight: 16,500 lbs.

Calf
Newborn length: 7–8.5 feet
Newborn weight: 265–353 lbs.

Drawing by Uko Gorter

CHAPTER *Four*

The Ancient Ones

I T WAS NATSILANE, THE TLINGIT SAY, WHO FIRST MADE KILLER WHALES. Natsilane was an ancient tribesman who faced a problem that remains ubiquitous even today: how to fit in with the family you marry into, especially your brothers-in-law.

The problem was that Natsilane was the finest hunter in the tribe into which he had married, and everyone knew it. Originally from another village, he had joined his wife's tribe to make her happy. However, his skills at hunting sea lions, a Tlingit meat staple, were widely renowned. He was known to be the first to leap from the canoes onto the rocks where the big brutes would haul themselves out, and he often accounted for half or more of the tribe's haul by himself.

Unfortunately for Natsilane, his wife's brothers had widely been considered the tribe's best hunters before his arrival, and they did not like playing second fiddle now. They would taunt him for trying to show them up. After years of building pride, they eventually schemed of a way to rid themselves of their annoyingly skilled brother-in-law. Out hunting in their canoes one day, they let him make that great leap on his own. Then, they steered away from the rocks where Natsilane now stood, after which they simply paddled away, telling those in the canoe who protested that winds had made it too dangerous to rescue him. Thus abandoned in the middle of the sea, Natsilane was approached by a sea lion who told him the sea lion chief wished to speak to him. The creature took him to the realm of the sea lion people under the sea. While there, he removed his own spear from the son of the sea lion king and healed the child, whereupon he was returned magically to his village, imbued with vast supernatural powers. He kept his return quiet and swore his wife to secrecy, then set about wreaking his revenge.

Natsilane's vengeance took the form of the carvings he made in the great trees he found near the shore where he went to work. One can readily imagine they looked rather like the great totem carvings of killer whales we see on Northwest poles today, except that when Natsilane put them in the water, they came alive and became killer whales. Or at least they tried; the first two carvings, made from spruce, failed, but the third, made of yellow cedar, became a great orca unlike any ever seen. Natsilane ordered it out to sea.

His murderous brothers-in-law were out hunting that day when they encountered the great beast, who destroyed their great canoe and then drowned the conspirators by preventing them from reaching shore. However, when the creature returned, Natsilane commanded it to refrain from such violence in the future and to act as a friend to men, assisting them where it might, and indeed, the great beast was seen from shore from time to time. It would bring freshly killed seal or halibut and leave it on the shore, and a feast would follow. In later years, more of the blackfish would appear. The legend holds that Natsilane was seen in his later years, riding pairs of the blackfish, a foot on the back of each: orca skiing, as it were.

There are some variations of this myth. In one, Natsilane is saved not by the sea lion king but by a mystical sea otter, who gives him seeds that grow various new trees, notably yellow cedar, all of which become important parts of Tlingit culture. Remember, it is from a carved yellow cedar tree that the first killer whale sprang alive.

The legends told by the Tlingits, who mostly lived in what is now southeastern Alaska and coastal British Columbia, stand in contrast to the mythology of many other coastal tribes (although the Haida tell a story of an abandoned hunter who gets revenge in much the same fashion as Natsilane), wherein killer whales—or "blackfish," *kakawin*, *Max'inuxw*, or a number of different names for orcas—are seen as predecessors to humans, not as creations of humans. Just as the Kwakwaka'wakw of western British Columbia believed that the first humans were killer whale spirits who emerged from the sea and transformed into men, the S'klallam who occupied the northern Olympic Peninsula likewise considered killer whales the ancient equivalent of humans.

A Southeast-Alaskan Tsimshian myth tells another version of the orca creation story. In this telling, the narrator takes on the persona of a transformed goddess:

> I used to be a beautiful white wolf; Noo Halidzoks created only one of me because she thought I was the most beautiful of the four-leggeds. I wandered throughout the world looking for someone like myself because I was really lonely. I came back very unhappy because my quest confirmed I was the only white wolf. I fell into a deep sleep. I had a terrific vision and when I awoke, I rushed to find Noo Halidzoks. "Ts'i'is (Grandmother), I have had a great vision; it calls for me to go below the great waters and sing the history of the world. I am so very lonely in my present form." Noo Halidzoks was sad, but realized the power of my vision and turned me into a shiny black whale. To remind me that I used to live on land as a wolf, she painted the white markings on my sides. I sank below the great waters singing my song, and all the other swimming creatures rushed to greet me as their new family member.
>
> I am also held in great awe for my power and size; it was believed I could capture a canoe and take it underwater to transform the occupants into whales. Thus, even to this day, a whale near the shore is considered to be a human transformed and trying to communicate with his family.

To all of these tribes, orcas are ancient creatures with immense spiritual powers. Orcas are the rulers of the ocean and the embodiment of its power. As it turns out, modern evolutionary science generally agrees with this.

• • •

Killer whales, the scientists tell us, have been around at least six million years and probably longer. As with all marine mammals, they are ultimately descended from land mammals. Native myths often associated killer whales with wolves; several myths describe them as wolves who transform into sea creatures. In evolutionary terms, however, there actually is little connection.

One of the land ancestors of all whales is a cloven-hoofed dog-like carnivore who lived during the Paleocene and Eocene periods, finally becoming extinct in the early Oligocene (that is, between about 62 and 30 million years ago) called the Mesonychid. Despite its appearance, however, it was not an ancestor of modern wolves. One of its evolutionary offshoots, called artiodactyls, were even-toed, hoofed creatures whose descendants would come to include pigs, cows, deer, and hippopotamuses. These last are considered the land mammals closest in relation to modern whales.

An offshoot of the artiodactyls was the first proto-whale, called the *Pakicetus*, although it did not look remotely like a whale. It was a furry little dog-sized creature, a carnivore that wandered the swampy coastline of the old Indian subcontinent, near modern-day Pakistan, during the early-to-middle Eocene period, between about 50 to 49 million years ago. The *Pakicetus* evidently hunted smaller creatures that lived in the swamp, including both fish and land creatures alike. It was a lousy swimmer, with heavy, compact bones that it used for ballast, and it was also not a fast runner. However, it was well adapted to a swamp life that included bottom wading, paddling, and swimming by undulating its body.

It was this latter trait that it passed on to the next descendant in the line to modern-day whales, a creature that could both swim and walk called the *Ambulocetus*. This early cetacean mostly propelled itself through the water by undulating its back in the manner of an otter or seal, but it also had legs and paws and could get about, albeit slowly, on land. Later, during the middle Eocene period, 49 to 48 million years ago, as did its ancestor, it mostly dwelled in the area of the Indian subcontinent (which at the time was a large island continent in the middle of the Indian Ocean). It also had an adaptation in its nose that allowed it to swallow underwater.

Next came the *Protocetidea* (proto whales), who propelled themselves through the water with a combination of undulation and strokes from their flukes, as modern whales do. The *Remingtonocetus* followed; they had developed more streamlined skull features and powerful jaws and apparently were fast and powerful swimmers. They also developed early hearing faculties for underwater listening. By now, moreover, these creatures were reaching more of the world's oceans.

One of the interesting components of whale evolution involved the way their breathing apparatus gradually moved from the front of the snout, as it was in the Pakicetes and Ambulocetes, as well as most land mammals, to the back of the head, a blowhole at the rear of the skull. In each step of their evolution, you can see the nasal passages moving farther and farther back, until they reached the skull's rear in early toothed whales.

Some in the line were strange-looking, none more so than the *Basilosauridae*, who appeared in the middle-to-late Eocene periods (about 38 million to 40 million years ago) and were found in all the world's oceans. They were probably the first fully aquatic cetaceans. They were also extraordinarily long creatures, with spines that extended fifty feet and longer, while their limbs shrank to just the tiniest vestigial rear feet and small forelimbs near the head that scientists speculate were used primarily for grasping partners during copulation. As the largest carnivore in the ocean, they had no predators.

The fossils of the *Basilosaurus* so impressed Herman Melville (who called them "the most wonderful of all cetacean relics") that he devoted a chapter of *Moby Dick* to describing a specimen found in Alabama in 1842 that was seventy feet long, a find that, as Melville notes, led to the somewhat mistaken naming for the creature: "The Alabama doctors declared it a huge reptile," he writes, "and bestowed upon it the name of Basilosaurus. But some specimen bones of it being taken across the sea to Owen, the English Anatomist, it turned out that this alleged reptile was a whale, though of a departed species."

Owen, Melville reports, eventually "pronounced it, in substance, one of the most extraordinary creatures which the mutations of the globe have blotted out of existence."

It was from these creatures that both branches of the modern whale family evolved. *Odontocetes* (toothed whales) are the most clearly descended from the *Basilosaurus*, with early forms appearing in the middle Oligocene through the middle Miocene periods (25 million to 11 million years ago). These creatures, called *Squalodon*, displayed the first appearance of echolocation faculties, although they appear to have been fairly crude. These were the direct ancestors of the first oceanic dolphins.

Baleen whales, called mysticetes, would appear about this same time, including the earliest, the *Cetotehriidae*. These creatures, who had both teeth for grinding and early hairy structures lining their mouths that were the forerunners of the thick baleen plates we know today, may have developed these features in response to some worldwide change in the oceanic environment and eventually went extinct. Other mysticetes, including the *Balaenidae*, the family of today's bowhead and right whales, appeared a little later, about 22 million years ago. They completely shed the teeth and developed thicker baleen structures. Eventually four distinct families of mysticetes would emerge. None of them appear to display echolocation capabilities, which are unnecessary, since their diet is comprised of food that floats or is at the ocean floor and does not require hunting or prey selection. (Worth noting: these whales do produce long-range calls at extremely low frequencies that could be a form of echolocation.)

The toothed whales would prove both more diverse and tend toward smaller sizes; nearly all of the great whales are baleen whales. However, that was not true in all cases. The earliest odontocetes were the *Physterids*, the early sperm whales, who appeared about 25 million years ago (about the same time as the earliest mysticetes) and grew to immense sizes. They, too, were deep divers, like their modern descendants.

Just as the Squalodons and other ancient dolphins were going extinct about 11 million years ago, in the middle Miocene, an explosion of other toothed whales, including the belugas, narwhals, beaked whales, and porpoises, emerged. The dolphins, part of the same shift, made the biggest splash of all.

Delphinidae are the largest, most diverse and most widespread of all the cetaceans, representing 36 species currently surviving, including dolphins, pilot whales, and killer whales. All of them possess some form of echolocation capacity, and all are carnivorous predators. Early forms of killer whales, scientists believe, evolved fairly early in the process, about 10 million to eight million years ago. The earliest fossil records of *Orcinus orca* date back about five and a half million years, suggesting they have been around at least six million years and probably longer.

In those six million years, orcas have been largely undisputed atop the oceanic food chain. They have been phenomenally successful over all those eons; they reside in every ocean on the planet and are the apex predator in each one of them. They have no predators themselves.

All of this led scientists originally to assume that all of the world's orcas were, in fact, a single species occupying different niches of different environments around the world.

That, as it turned out, was all wrong.

• • •

Thanks to a Canadian whale researcher named Michael Bigg, sometime in the 1970s scientists first began realizing that there might be what they call "speciation," the divvying up into individual species, going on among killer whales. It was Bigg who had pioneered the use of photo identification, focusing on the dorsal fins and white "saddle patches" that are the most frequently visible parts of killer whales, and which act as a kind of unique fingerprint, to take a census of Pacific Northwest killer whales. With help from a number of colleagues on both sides of the border (notably, Paul Spong of OrcaLab, and John Ford and Graeme Ellis of the Department of Fisheries and Oceans on the Canadian side, and Ken Balcomb of the Center for Whale Research and Rich Osborne of the Whale Museum on the American side), Bigg was able to get a fairly clear picture of the orcas' grim population outlook within a few short years.

The majority of the orcas that Bigg and his colleagues observed were what they called "resident" killer whales, individuals who spent a great deal of time in the inland waters of the Salish Sea and northern Vancouver Island. These whales moved together in large pods, sometimes even convening in gigantic "superpods" of fifty whales and more, and they fed strictly on fish.

However, Bigg also observed other whales who behaved quite differently, moving in small pods of three to five (or even less, sometimes), swimming not in the deeps where the fish were but in the shallows, where the seals and sea lions upon which they mostly fed could be found. (One of Bigg's earliest observations of wild killer whales, in fact, involved a pod of whales attacking and eating a Steller sea lion.) At first he thought these were renegade or outcast whales; however, the more he and his colleagues watched them, it became clear they were actually a separate population, a different kind of killer whale than the residents.

They named them "transients," and indeed these were whales who would roam extraordinary distances, from the coast of California to the Queen Charlotte Islands (or Haida Gwaii), and everywhere in between, including the Salish Sea. They looked different: The females' dorsal fins were more triangular and pointed in appearance; their saddle patches were a solid, uniform grey; and the whales in general were often more nicked up and scratched. In addition, they did not vocalize heavily while hunting, unlike the residents, relying instead on stealth to capture their prey—and what a range of prey it was. Over the years, scientists would observe these whales devouring all kinds of mammals: seals, sea lions, harbor porpoises, Dall's porpoises, Pacific white-sided dolphins, and even a few varieties of seabird. On a couple of occasions, they were even observed taking down and eating the moose that would sometimes swim between islands in the archipelagoes of southeast Alaska and British Columbia.

Unlike their salmon-eating counterparts, the transients were not warm and cuddly sea creatures to observe but were brutally efficient killers who could tear apart a harbor seal (which comprised over 60 percent of their diet) in seconds. Their role in the ecosystem was similarly indisputable.

Over the fall of 2002, articles began to appear in the regional press describing biologists' concerns about a sudden overpopulation of harbor seals in Hood Canal, one of the more remote and traditionally pristine reaches of southern Puget Sound. They were seeing upwards of 1,500 seals in the canal, putting stress on the local fish and mollusk populations. Then, early in 2003, a large group of 11 transient whales showed up in Hood Canal and began devouring the seals for weeks on end. It was somewhat unprecedented, since transients rarely ever stay longer in a single place than a day or two at most. These whales stayed in Hood Canal two months and from nearby shores could be seen frolicking with their prey, tossing them in the air repeatedly, like a cat tossing a mouse it has caught. After the first seven weeks, scientists estimated that the whales had devoured about a third of the seal population in the fifty-mile-long fjord.

Then the whales disappeared, although in 2005, after the seals rebounded to number about 1,200 in Hood Canal, another group of six transients took up occupancy there, this time for over five months. Dubbed "the slippery six" by locals, they too spent plenty of time feasting on harbor seals. The seal overpopulation problem went away—scientists estimated the whales ate half the population—and eventually so did the orcas.

"It's amazing to see them hunt," a state biologist named Steve Jeffries told a reporter. "They're able to kill seals without any effort at all. They'll be swimming along, and then they just go down and pick up a seal."

Watching transients hunt can be simultaneously appalling and enthralling. Among their prey are Dall's porpoises, which are among the fastest animals in the water where they can reach 35 mph. However, orcas, even more remarkably, are perfectly capable of keeping up with them, despite their significantly greater size, mainly by "porpoising," thrusting themselves powerfully forward through the air in a series of linked leaps. There is no scene quite like watching a transient killer whale chasing a Dall's porpoise, sprinting high out of the water at full speed as though it was born to fly.

The more scientists studied them, the clearer it became that "transient" was not exactly an accurate description. These mammal-hunting whales, it appeared, formed fairly large but broadly dispersed communities that ranged over wide stretches of territory with an ultimately well-defined reach. These orcas formed smaller pods but, within their own communities, were also remarkably cooperative.

Nor did they appear to socialize with resident killer whales at all. In the intervening years, all of the interactions ever observed between resident and transient killer whales were of a hostile nature, with the residents —who always outnumbered the transients—chasing their mammal-eating neighbors away. Scientists aren't sure why, but they speculate that perhaps at some level the transients could pose a threat to the residents' young calves. Not only was there no social overlap between the populations, but there was no communication between them, either. Resident orcas used an entirely different set of calls from transients when communicating and used them in entirely different fashions. It was a biologically unusual situation for two populations of the same species to occupy the same space (or waters) and to observe completely different behaviors, including diet and communication. The only word for this is *culture*, and there is only one other species that has exhibited it: human beings.

About the only thing the resident and transients had in common was their echolocation. Scientists began to wonder if there was even any genetic connection. Finally, in 2003, the genetics results came in, and the answer was definitive. There had been no genetic interaction between resident and transient killer whales for somewhere between 150,000 and 700,000 years. To put that in perspective: *Homo sapiens* has been on the planet only about 200,000 years. That in turn raised all kinds of questions: Were residents and transients separate species? What about the other populations of killer whales around the world? Is there a single species of killer whale or a number of them? That debate, in fact, still rages today. However, the fact that it exists at all is testament to the legacy of Michael Bigg, who died of cancer in 1990. Indeed, today the preference among scientists is to stop calling these orcas "transients." Their new name: Bigg's killer whales.

• • •

Contributing to the debate was the realization, in the mid-1980s, that there was still a third population of killer whales in the North Pacific, dubbed "offshores," who lived much farther out to sea and never came in to shore. These whales, too, had a distinctive appearance and diet; their coloration was slightly different; they appeared to be smaller; their fin shape was also different than that of transients or residents; and they appeared to be eating deep-diving sharks. They also communicated with a completely distinct set of calls. Given their known location, however, these whales have been notoriously difficult to study, and so relatively little remains known about them, including the size of the population. All we know for certain is that they exist. Thus we know there are at least three distinctive populations of killer whales in the North Pacific, all in overlapping waters and territories.

However, these are hardly the only orca populations around the world with distinctive diets and cultures. In the North Atlantic, at least two different kinds of orcas have been observed, divided again between fish eaters and mammal eaters, while Argentina orca populations have been observed eating young leopard seals and fish alike, and in the Antarctic, at least two different kinds of killer whales have been observed devouring nearly everything that moves.

By 2011, a general consensus began to take shape in the scientific community that studied killer whales: *Orcinus orca* might best be understood as a "species complex," an umbrella term for a species, disparate populations of which technically can interbreed (sperm from a male North Atlantic killer whale, as we have learned from orcas in captivity, can impregnate a Northern Resident female), but who choose not to for various reasons that are purely cultural, setting up disparate "ecotypes" with distinctive behaviors, communications, diets, and cultures.

The debate focused around a proposal by geneticist Philip Morin, who examined DNA samples from a variety of killer whale populations, that there were three new species of killer whales, along with a number of subspecies. Morin's study found that North Pacific, North Atlantic, and Southern Ocean killer whales represented three species that were "independently evolving lineages" and deserved species status. Within those three, he suggested that other ecotypes such as mammal-eaters and fish-eaters be designated subspecies, pending further study on their evolutionary separation.

Longtime orca researcher Lance Barrett-Lennard of Vancouver suggested the "species complex" concept as a middle course, based on what he had observed about them, particularly their "social exclusivity," which he explained "predisposes whales to form diverse, genetically isolated populations—incipient species, effectively."

If the whales survive, he says, "we could be lucky enough to be witnessing the early stages of an adaptive radiation of killer whales whereby a variety of new species will exploit diverse ecological niches—or we could be looking at an ongoing process by which new ecotypes form and periodically wink out.

"If I'm correct and killer whales are in the relatively early stages of an adaptive radiation, the populations we see at present represent a continuum of continually diversifying forms," and these, he says, could easily develop into a fully independent species.

So far, there are ten different ecotypes of killer whales that scientists recognize under the "species complex" approach, divided by Northern and Southern hemispheres.

Northern:
Resident Killer Whale
Almost certainly the best-studied wild cetaceans in the world—not just in the Pacific Northwest, but in Alaska as well—the more we learn about these highly social animals, the more we realize how little we actually know. All told, there are over 900 of these animals in the North Pacific, including about 80 Southern Residents from the Salish Sea, 200 Northern Residents from northern Vancouver Island, about 500 resident whales in the Gulf of Alaska, and another 100 or so elsewhere in southeastern Alaska. Highly stable social creatures, their communications are sophisticated, gregarious, and enduringly mysterious. Their fish-eating diet renders them vulnerable to overfishing of stocks by humans, and their fearless and simultaneously friendly dispositions (not to mention their relative predictability) make them a favorite of tourists and whale-

watching operations. They favor pods of five to twelve members, but they also frequently gather socially with pods from within their larger community groups, sometimes even forming large "super-pods." Genetically speaking, they are also one of the most distant relatives of the whales with whom they share their waters.

Bigg's Killer Whale

They are also known as "transient" orcas. Even though they occupy the same waters as resident whales, they have never been observed socializing with them. Rather the opposite; whenever the two happen to coincide in the same area, the reaction uniformly has been that the residents drive the Bigg's whales away. Bigg's whales primarily eat harbor seals (over 60 percent of their diet), but they pretty much devour everything within reach: sea lions, harbor and Dall's porpoises, even seabirds and squid (not to mention the odd moose). Their social structure is matriarchal but is considerably more fluid than that of resident pods, and they tend toward smaller pod sizes. They also are less inclined to large social gatherings. Scientists estimate that there are about 350 of these whales in the North Pacific.

Offshore Killer Whale

The existence of these whales, who live in the open sea of the North Pacific and travel as far south as southern California and as far north as Kodiak, Alaska, was first noted by whale-watch tourists in 1988 and confirmed in 1990, but the difficulty in studying them (the waters they occupy are among some of the most notoriously rough if not outright lethal) leaves scientists with little information. They are physically smaller than both residents and transients, and their saddle patches are very faint, making them tricky to ID. Scientists have been able to identify over 200 of these whales, but many more have not been photographed. Based on documented predation, they are believed to feed primarily on deep-diving Pacific sleeper sharks. Three offshore killer whales who washed up dead along the coast of western British Columbia were found to have teeth worn down almost to the gums, something never seen in other Pacific orcas, a sign they had been feeding on sharks with their abrasive, sandpaper-like skin.

Eastern North Atlantic Killer Whale, Type 1

These whales are notably smaller than their Type 2 cousins (as well as most Pacific orcas), and scientists have puzzled over the worn teeth that typify these orcas; the speculation is that since a large portion of these killer whales' diet is small fish like herring and mackerel, they wind up grinding their teeth more than their mammal-eating relatives. It's also possible that, like offshores, they are dining on sharks. However, these whales have been known to eat marine mammals, also, notably harbor seals. They swim mostly in the waters around Norway where they number around 700 and Iceland where there are about 400 of them. They also appear off the northern coast of Scotland.

Eastern North Atlantic Killer Whale, Type 2

These whales mostly occupy the waters around Ireland and Scotland and number about 400. They are larger whales but feed exclusively on marine mammals, including dolphins, porpoises, and baleen whales, especially minke whales. This is a much smaller population of orcas, but they are morphologically quite distinct from Type 1 whales, being genetically closer to Antarctic whales than to their neighbors.

Southern

Antarctic Type A Killer Whale

The largest and best known of the Antarctic orcas, these animals migrate south during the summer to feed on minke whales and elephant seals along the shore of the continent. During the winter, they migrate northward to warmer climates, including some parts of the tropics. These orcas make appearances around New Zealand as well. They grow to over thirty feet in length and have distinctive saddle patches, and like other orca populations, appear to have a matriarchal social organization. It is also the largest orca population in the world, numbering around 15,000. (All told, the four Antarctic orca populations comprise over 70 percent of the world's estimated total population of 100,000 orcas; one survey put the total Antarctic population at 80,000.)

Pack Ice Killer Whale

A slightly smaller version of the Antarctic orca with a different complexion, its skin is almost gray, with a darker section called a "dorsal cape" running from its rostrum to its fin. The white parts of its skin are slightly yellowish. These whales live around the continent, where they mostly forage for Weddell seals in loose pack ice; famously, they can be seen using formation hunting patterns to create waves that wash the seals off ice floes. They have also been observed training their young in these techniques, without actually making a kill but as practice. They also occasionally kill minke whales.

Gerlache Killer Whale

An even smaller version of the pack ice whale, it also has a grayish complexion in two tones, and its white skin also often has a yellowish tint due to being infested with algae. Its eye patch is smaller than that of its larger cousins, too. These orcas occupy the Gerlache Sea and the Antarctic Peninsula, where they appear to feed mostly on penguins.

Ross Sea Killer Whale

This is the smallest of all the killer whales, with adult males reaching only about 20 feet in length. Like its Antarctic cousins, it is also slightly grayish with a cape and yellowish skin from collecting algae. Its eye patch is distinctively small, narrow, and slanted. It is believed to feed primarily on fish and can be commonly seen in the sea for which it is named, as well as in the pack ice of the eastern Antarctic.

Subantarctic (Crozet) Killer Whales

These are little-studied whales who live outside the Antarctic and primarily feed on fish in the waters between the Indian Ocean and the Antarctic. They are best known for stealing fish from the lines of fishermen who work in these waters (leading to frequent conflict and the deaths of a number of whales). However, these whales appear to be omnivorous, since they have been seen eating not just fish but also penquins, seals, fur seals, and other cetaceans.

These really are only the best known and documented populations of killer whales. The complete list of populations includes the killer whales who inhabit New Zealand's waters much of the year, feeding primarily on rays and sharks and numbering about 300; rare, white orcas have been observed among a population of some 700 killer whales off the coast of eastern Russia's Kamchatka region; populations off the coast of Argentina, the Caribbean Islands, the Galapagos Islands, Japan, the Strait of Gibraltar, Australia, and Tasmania; and small populations in places like Papua New Guinea, the Chagos Islands, Northwest Scotland, the Falkland Islands, and the Subantarctic Prince Edward Islands, off the coast of South Africa.

Recent genetic studies of mitochondrial DNA from all these various populations have demonstrated that the genetic relations among them vary widely. Resident and offshore killer whales in the North Pacific are relatively closely related, while transients in the same waters come from one of the earliest breaks in killer-whale lineage, some 700,000 years ago. Those North Pacific residents are from the same genetic grouping as North Atlantic whales, at least the fish eating Type 1 whales. Type 2s are from the same genetic grouping as Antarctic whales, while Bigg's whales are off in their own universe with Ross Sea killer whales and some Eastern Tropical Pacific populations.

What the genetic map demonstrates, clearly, is that killer whales have been gradually diverging into separate populations that are so culturally distinct that they have become biologically distinct as well; if not speciation exactly, they are a dramatic living example of the process.

ORCA SOCIETIES

What is remarkable about these populations is that they also have distinctive social organizations—variations on a larger theme, to be sure, but each suited to the vicissitudes of its environments and prey types.

The larger theme is that orca societies are matriarchal. In each of the groups studied at length (and this does exclude some populations, notably the Subantarctic killer whales), orca populations are structured around their mothers. The words "pod" and "matriline" are almost interchangeable when it comes to orcas' social structure, but there are variations on this theme.

In the classic structure from the best-known orca societies, the Northern and Southern Resident orcas of the Pacific Northwest, killer whales of both sexes remain with their mothers for life. When the mothers die and leave behind two daughters with offspring, the daughters will then often split off into their own separate pods, while remaining associated generally during larger social occasions. Regardless, the males remain with their mothers. In the event of her death, male residents will sometimes disperse to other pods with which they were socially attached previously.

Thus, in the J pod of the Southern Residents, orca watchers could enjoy for many years the spectacle of J1 and J2, Ruffles and Granny, the two best-known whales, known for leading the J pod on its arrival at Lime Kiln and other whale-watching vantage points. They were immediately identifiable; he for the massive six-foot fin that looked, on the trailing edge, like a ruffled potato chip (wavy dorsals are fairly common among Southern Resident males), she for the little half-moon nick in the middle of her dorsal. They were also renowned for their ages: When J1 finally died in 2012, it was estimated that he was sixty years old. That same year, Granny, his mother, and the matriarch of the clan, turned 100. She was still alive as of 2015.

Likewise, the K pod whale known as Cappuccino (K21) was more or less orphaned in 2012 when his mother, Raggedy (K40), died suddenly. A big male, fond of spyhopping kayakers and wowing the tourists

K-21, "Cappucino," spyhopping.

at Lime Kiln with his big breaches, he simply picked up his bags, so to speak, and took up permanent residence with the J14 pod, a different clan altogether but a group that had been seen in company of Raggedy and her band at various times over the years.

These are lifelong bonds and, indeed, appear to be essential to each killer whale's self-identity. Home, for these whales, is not a place. Their home is each other. This profoundly affects the behavior of the killer whales, because the pod's well-being is essential to their own. This is why, throughout killer whale societies, prey sharing is common. Cooperative behavior is the rule, and physical conflict is almost completely unknown. However, all this is only strictly true of the resident killer whales of the North Pacific, including those in southeastern Alaska and western British Columbia, which have a chief common trait of being strictly fish-eating orcas, with Chinook salmon comprising over 80 percent of their diet. Their other common trait is that they tend to socialize in larger groups of ten to twelve whales and sometimes in even larger superpods.

In other orca ecotypes, however, the details of the structures vary within the matriarchal theme. Among Bigg's killer whales, for example, the pods tend to be much smaller and less prone to socializing. This is probably an adaptation to the requirements of hunting larger marine mammals, when stealth is required (this is also almost certainly why Bigg's orcas are so disinclined to vocalize, at least until after they make a kill, at which time they are known to make a lot of distinctive sounds). A large group will have much more trouble sneaking up on seals and sea lions in their haul-outs, a job that can be handled more efficiently with just a few whales and a single seal can feed five, but not many more, whales.

If a Bigg's pod grows beyond five or so whales, it will often either divide into a couple of pods, or the extraneous mature offspring, both males and females, will wander off and join other pods or travel solo. The mother remains the main organizing principal of the pods, with males playing a secondary role, evidently as hunters for the larger mammals, such as sea lions. However, it is a much more fluid social order, with pod members mixing frequently with other transient pods. Unlike the residents, Bigg's orcas will not necessarily remain with their birth pod for life.

The best-documented case of this kind of dispersal of these whales involves some of the first mammal-eating whales encountered during the period when killer whales were being captured from Northwestern waters for display at marine parks. In 1970, six orcas, including a rare albino orca, were captured and held in a sea pen at Pedder Bay, near Victoria, the provincial capital on the southern end of Vancouver Island, British Columbia. At the time, orca captors had previously only had successful captures with fish-eating killer whales, and so they were mystified when these six whales refused to eat the fish they were given.

The albino whale, who was a female, and one of the young females, who they named Nootka, proved more amenable to the fish diet, and they were both soon moved to an aquarium display near Victoria called Sealand. The other three whales remained in the sea pen for several months: a large male with a jaw deformity who was given the name Charlie Chin and, later, the scientific designation T1, for Transient 1; his mother, designated T2; and another adult female, designated T3. T3 was the mother of the albino female, who was numbered T4, and was named Chimo. Her young female companion, Nootka—probably her sister—was designated T5.

The three who remained behind refused to eat for weeks, and after 75 days, T3 died of malnutrition. Finally, four days later, T2 began to eat fish, and Charlie Chin soon followed. It wasn't a diet they understood (no doubt it did not occur to their captors to toss a seal into their pen), but they adapted.

Things continued this way for another four and a half months, until one night someone threw a set of weights over the top of the net that held the whales in at Pedder Bay. The two survivors simply swam out of their prison that night and resumed their normal lives.

The two who remained in captivity in Sealand were not so fortunate. Chimo died two years later at the Victoria Aquarium of a lung infection; it was later ascertained that her albinism was a product of a genetic disorder that also made her more prone to illness. Nootka lived a good deal longer, surviving as a performing orca and being shuttled from one aquarium to another, with stays in Ontario, California, and Texas, before finally settling in for a stay at Sea World in San Diego, where she finally died in 1990, of a kind of pneumonia associated with ulcerated infections that turned her lungs into a mass of inflamed tissue.

The two escapees picked up where their lives had been broken. This was especially true of T2, named Florencia by her captors. She and her son, the easily identified Charlie Chin (who also had a large notch on his dorsal), were observed in Northwest waters frequently over the ensuing years, and in short order she had given birth to another male calf, designated T2a. Six years later, in 1979, she gave birth to a female calf designated T2b, at which time T2a mysteriously disappeared.

This pod continued to be one of the most commonly encountered families of Bigg's whales throughout the 1980s in the Salish Sea, and their activities became well documented. It appeared to be an extremely stable family group. Then, in 1986, Charlie Chin started taking off from his pod and was seen swimming alone in distant reaches, sometimes as far north as the Haidi Gwaii, foraging solo. It was an unusual sight, and his striking appearance made it more so.

At the same time, scientists suddenly encountered T2a, who was found alone, hunting seabirds off the northern coast of Vancouver Island. He was seen a number of times over the next few years in the fjords of southeastern Alaska, the last time in 1988.

Charlie Chin, meanwhile, continued to reappear with his mother's pod from time to time. After T2 gave birth to another calf, designated T2c, over the winter of 1988-89, the big male returned to being a regular fixture of the pod. For several years, the four whales—T1, T2, T2b, and T2c—remained a stable unit and were even seen occasionally in the company of other transients, something that had not been common for them. However, in 1992, T2b began wandering off and was seen in the company of various other Bigg's pods, and Charlie Chin shortly afterward simply disappeared again. It's unknown whether he dispersed again or died, but the latter is suspected, as it has been so long since he was last seen.

T2 and T2c remained a traveling unit for many more years, frequently appearing with other pods. They were occasionally rejoined by T2b; she has tended to associate with a couple of other Bigg's pods in the ensuing years. T2 finally disappeared in 2011 and is now presumed dead, but her daughter, T2c, has had two calves and now has formed a pod of her own, currently comprising four whales. It is believed T2b has also given birth to a calf.

All this is in striking contrast to the social organization of resident pods, in which rigidity tends to rule. There is some occasional interpod drifting, as when Cappucino picked up with a J pod family, but this usually arises out of unusual circumstances, such as the death of a mother. On one occasion, when one of the K pod's matriarchs—Lummi (K7), a whale estimated to have been born in 1910—passed away in 2008, her own daughter, named Georgia (K11), believed to have been born in 1933, inherited the mantle as matriarch. Out of the blue, a strapping 16-year-old male from the L pod named Onyx (L87), whose own mother had died three years previously, "adopted" her, appearing as the elderly Georgia's rather imposing male escort for her remaining years, until she died in 2010. Since then, Onyx has traveled exclusively with J pod whales. (Go figure.)

Those, however, are the exceptions to the rule. Studies have consistently found that there is very little social drift among resident killer whales; once born into a family, they remain with it until they die. They are gregarious in their social dealings with other Southern Residents, but over the long term, they remain closely identified with their matri-line. However, there is a generous quality to all this. What the excep-tions tend to prove, in fact, is that while orca society is outwardly rigid, it is flexible enough to accommodate the special needs of members who undergo family hardships; resident whales who are orphaned never seem to want for a family they can call home.

Again, this varies from population to population, generally de-pending on prey types. Mammal-eating orcas appear to tend toward more fluid matrilines and less social rigidity, while fish-eating popula-tions tend toward lifelong familial bonds. Some populations, such as North Atlantic killer whales, seem to consume both kinds of prey and yet appear to trend more toward resident-like social rigidity. There are four identified types of Antarctic orcas, but all four appear to be flexible. However, most of these populations have been so little studied that, in fact, we know very little about their social organizations.

ORCA CULTURE

Two things really define orca culture: food and communication.

It is natural that food would be central to the killer whales' world. Unlike the great baleen whales, who are capable, without eating, of mi-grating thousands of miles on the blubber they carry and need only eat during key seasonal periods at their feeding grounds, killer whales have comparatively small fat stores and thus have a pretty near-constant need to eat. Their travels, their movements, even their sleep periods are de-termined by the need to eat; fish eaters go where the fish are, and mam-mal eaters go where the mammals (and seabirds) are.

However, for all of these creatures, for fish and mammal eaters alike, hunting is done creatively and collectively. Even if an orca manages to capture and kill its prey by itself (notably among fish eaters), it is still common for it to share its meal with other members of its pod, particularly the young. However, just as often, orcas hunt cooperatively:

- **Chasing Salmon:** Resident salmon-eating orca have been observed apparently driving salmon into underwater cliff walls, confusing the fish and making them easy pickings, even for young orcas. Adults have been observed apparently training calves in hunting methods using this technique and then sharing the catch afterward. At other times, orcas will be observed chasing salmon toward their waiting comrades, all then sharing the catch.
- **Herding Herring:** North Atlantic killer whales have been observed herding herring into balls as a group, a technique called "carousel feeding." After forming the balls, they slap the fish with their powerful flukes, stunning them and rendering them easy to scoop up by the mouthful, which the orcas then do with gusto.
- **Pinniped (Seal) Prey:** Bigg's whales have been observed using a variety of cooperative techniques when hunting marine mammals such as sea lions and seals, often handing prey off to one another and then joining the feast at the end en masse. At times, seals have been observed diving and hiding in rocks to escape killer whales, so their pursuers will take turns waiting underwater for the seal to emerge from the rocks, one coming up for air while the other watches and waits, making it a simple chase when the seal itself finally has to emerge for a breath of air.
- **Killing Whales:** Other mammal-eating killer whales have been frequently observed acting like wolf packs when pursuing and killing humpback whales, spending hours chasing the whales, exhausting them by taking turns at harassing them, until one or two of them finally make the kill, and then all the members of the pursuing pod take part in the ensuing meal, sometimes for several days thereafter.
- **Sting-Ray Teamwork:** In New Zealand, killer whales who feed off sting rays living near shore have been observed hunting cooperatively to avoid being stung by their prey. One whale will grab the ray in its mouth and flip it over, rendering it immobile, at which point its partner will swoop in and make the kill. Both orcas then share in the meal.

- **Beach Snatchers:** Along the shores of the Valdes Peninsula in Argentina, killer whales hunt for young elephant seals by beaching themselves and snatching their prey off the shoreline. Adults have been observed teaching young orcas this technique by grabbing seals and releasing them near the juveniles so that they can practice the difficult work of capturing and killing the prey without the risky beaching maneuver.
- **Ice-Floe Harmonics:** Perhaps the most impressive bit of coordinated hunting, frequently replayed on wildlife programs, is that used by Antarctic orcas when working to force seals they are hunting off ice floes. They do this by lining up and swimming as a team directly at the floe, creating a large wave that then washes the seal off the ice and into their waiting jaws.

Their food actually creates many of the defining lines of orca culture. Southern Resident killer whales, for instance, have been observed to be highly selective in their diets. In the summertime, it seems, they not only feed exclusively on Chinook, but appear to insist on eating only Chinook from the Fraser River in British Columbia. Samples taken from their foraging and excrement indicate that when the orcas are in the interior waters of the Salish Sea, they are only eating Fraser Chinook in spite of the presence of abundant Chinook from other watersheds.

This is only one aspect of the broader "social exclusivity" that Lance Barrett-Lennard describes in orca populations and that may well be the driving force in the hardening of these broad-ranging ecotypes into disparate subspecies among the world's orca populations. It is driven largely by what some orca observers call a kind of "cultural rigidity" that informs nearly every aspect of killer whales' lives.

"Orcas' behavior is almost totally governed by cultural conditions and hardly at all by what we were taught in school about instincts and stimulus response and animal behavior in general," says Howard Garrett of the Orca Network, himself a trained sociologist. "It really depends on what they've learned and what they're supposed to do, what is expected of them within their societies and cultures. This culture is handed down, passed down over the generations, and modified and developed over time.

"But even if it means their demise, even if it means they starve, even if it means they become susceptible to capture teams—as we saw with the orcas at Penn Cove in 1970—they will maintain their absolute cohesion. And it is the same with the rest of their culture: their diet, association patterns, calls."

This latter is the communication component of orca societies. Each distinct population has its own "dialect," its own set of stereotyped calls that it uses when communicating with fellow pod members. The immediate linguistic function of these calls is not entirely known, although in one study of the structure of the calls, it was concluded that many of the sequences of the calls relayed broad motivational information and that certain sub-classes of vocalizations apparently contain "more subtle information on emotional states during socializing."

However, there are also calls that killer whales use repeatedly in speaking to each other, the immediate function of which appears to be a kind of beacon, so that the whales know the precise location of each other while hunting and just traveling generally, although these calls also appear to contain information indicating pod identity. Some of these calls are used across the entire clan of Southern or Northern Residents, while others are specific to individual matrilines and appear to indicate those identities.

Canadian whale scientist John K.B. Ford pioneered much of the study of orca dialects by studying the vocalizations of Northern Residents for many years. Ford established that certain resident dialect calls could indicate community identity, while others indicated individual pod identity, and that subtle changes in these intrapod calls often occurred over time. Even more important, these changes were transmitted not only within the individual pods but also across the broader Northern Resident community, establishing that cultural transmission occurred not only vertically, from mothers down to their offspring, but also horizontally, between disparate pods within the broader clan.

As scientists Hal Whitehead and Luke Rendell explained in their definitive 2001 study, "The complex and stable vocal and behavioural cultures of sympatric groups of killer whales appear to have no parallel outside humans and represent an independent evolution of cultural faculties."

Of course, it is important not to draw conclusions that are too broad from the realization that killer whales have relatively sophisticated societies and a comparatively complex culture—compared, that is, to other animal societies. Compared to human society, on the other hand, it appears to be fairly simple. Similarly, the surface simplicity of the killer whales' communications, as Justin Gregg would argue, does not appear to qualify these communications for the term "language" as we know it, since it is not clear at all that there is any great complexity to them.

However, that is more a function of what we do not know about killer whales than of what we do know; what we know about their communications, as well as the large and complex brains behind them, certainly leaves open the possibility, if not the likelihood, that there is more going on there than meets the eye or ear.

Likewise, orca culture, to all outward appearances (and compared to human culture), seems relatively simple and unsophisticated. However, it is worth remembering, too, that it is unique. While other animals—notably chimpanzees and other primates, as well as certain species of birds—exhibit a number of cultural traits, in no other animal, besides humans, are there such distinct, definable, and variable cultures as those that can be found among all killer whales.

While we are at it, it is worth comparing human culture to orca culture, because the contrast is deeply revealing. "The most striking aspect of orca culture, to me, is the sheer absence of internal strife and the predominance of cooperation," says Howard Garrett. "They have hierarchies of dominance, beginning with the matriarchs, but there's no sign of discipline, there's no jousting for position. You only see occasional scuff marks or rakes, mostly on young ones. But they don't butt heads; they don't beat each other up.

"Then, one of the few universal behaviors that they have is that they do not harm humans. They are unique among apex predators that way. So it's clear that even though they are capable of extreme forms of aggression—just ask their prey—the prevailing ethos of their culture keeps them from harming each other and from harming other life forms they choose not to eat. It's actually a remarkable display of discipline and intelligence.

"And to note that humans could learn from that is probably an understatement."

CHAPTER *Five*

The Demon From Hell

FOUR SISTERS ARE OUT WALKING ON THE BEACH. IT IS A ROCKY BEACH, still wet from the receded tide. There are barnacles everywhere, and the kelp pods crackle and pop beneath their feet as they amble through it. They look for oysters to collect in their cedar baskets. It is a clear summer morning, and the water is calm like glass. Suddenly the stillness is broken by the sound of a killer whale, the *Max'inuxw*, as it bursts above the surface—*kooosh!*—and sends its bushy plume skyward, where it lingers in the morning air. The sisters gasp and point.

"The ancient ones!" they say. "A good omen! Look how beautiful!" The oldest sister is entranced. The pod of whales is led by a big male with a six-foot dorsal fin. Seeing that she is watching him closely, the orca approaches and sticks his head above the water, spyhopping.

The male orca A-13, cruising the shore of northern Vancouver Island.

"I am going to go talk to them!" she tells her sisters. "Perhaps they'll give me a ride."

"No don't!" they exclaim. "It's dangerous!"

"Pssshhh!" she replies and walks down to the water's edge. She beckons to the big whale.

"*Mi'Max'inuxw*, come on in. I want to take a ride!"

The next thing she knows, he is there standing next to her on the beach—a large man with dark skin. He has a large canoe.

"Do you know how to use this?" he asks her, gesturing to the canoe.

"I am willing to try," she answers, her chin out bravely.

He pushes the canoe, with the sister, back into the water, and he becomes the whale again. Now he opens his mouth wide, wider than she has ever seen, and tells her to step in. She does. Inside him, it is a large space, and he tells her to go to the back and use the fins and flukes to steer. She does so, and they set out to sea, swimming and frolicking around the bay.

They come back to shore and he opens his mouth wide again and she steps out, smiling broadly. "Amazing!" she tells her sisters, who are agape.

The orca asks her: "What about your sisters? Would they like to try?"

"No way!" cry the two middle sisters. "Nuh-uh! You wouldn't catch me trying that!"

But the youngest sister steps up and says she'd like to try. So the orca comes up to the shore and opens his vast mouth; she steps inside and goes to the back to work the fins and flukes. They sail out for a ride, splashing and laughing around the bay. The orca comes back in and opens his mouth again, and the youngest sister comes out laughing and giggling.

"No one else?" the orca asks.

The two middle sisters are still too frightened to try, but the oldest sister steps up and says: "Well, if they're too scared, I'd like to go again!"

So the orca opens his mouth; she steps inside and they swim off, but this time he does not come back.

After a while, the three remaining sisters realize that the orca is not returning with their sister. They run back to the village and tell their father, the chief, what has happened. A search party is formed, and the whole village spends the next eight days searching for the missing princess. After many more days, they finally give up and mourn her as lost.

A few weeks after this, the youngest sister is out by herself collecting abalone in her cedar basket. Again she hears the sound of the *Max'inuxw* and turns to look, and there is her oldest sister standing on the shore. A piece of seaweed is draped across her brow.

"Hello, little sister!"

"Big sister! We have been so worried! We thought you were lost!"

"Tell father and mother not to worry. I am married now to *Mi'-Max'inuxw.*"

"Are you all right? Are you happy?"

"I am happy," she says. "He is king under the sea."

With that, she returns to the water's edge and climbs into the vast mouth of her husband, and they disappear under the sea. And that is the last she is ever seen—by humans, that is.

• • •

That is how the Kwakwaka'wakw tell it, anyway. There are several versions of this Northwest Coastal legend. In the telling of the S'klallam tribe, the princess Kakantu tells her sisters that she admires the whales: "That one has a handsome face," she says, pointing to the king. "I wish he were my husband." That night, her father is visited by the blackfish king in his human guise, and he bargains for the daughter's hand in marriage, offering an ever-abundant supply of fish in exchange for her. A slave girl tries to substitute herself for the daughter but is caught and called out, and the daughter gladly goes to live with her new husband.

The blackfish king brings his wife back to her village periodically to see her family, as he promised in his bargain with her father. However, after a few of these visits, her mother notices the seaweed growing from Kakantu's hair and the pallor of her face and tells her not to come back: "You've become a supernatural thing," she says, "so whenever it was to be that you go, then go!"

What's noteworthy about these legends—and for that matter, nearly all of the Northwest Coastal legends pertaining to the killer whale—is the full personage that killer whales are accorded. They transform into men when they come to land, return to their whale form when they go in the water, and emerge back in their own villages in their human form again. Certainly, it is not unacceptable for the daughter of a chief to marry such a man and doing so brings great power, spiritual and material, to the tribe in return.

There were other cultures that similarly honored killer whales. The native inhabitants of Newfoundland, the Maritime Archaic cultural complex, honored orcas as fellow hunters and made ornate stone carvings of them that were later uncovered by archaeologists. The same is true of the Maori people of New Zealand, who called orcas *upokohue*. In Japan, killer whales were an ingrained part of the maritime folklore of this island nation. Their name in Japanese is *shachi*; rooflines of Japanese homes and ancient castles are often decorated with figures called *shachihoko*, half fish, half dragon (or tiger), that is a kind of caricature of a *shachi*.

The Japanese, dating back to at least the 7th century, were also among the first whalers. They mainly hunted right whales, humpbacks, fin whales, and gray whales, but they occasionally harvested blue and sperm whales. After all these stocks became depleted in the 20th century, however, the Japanese took to hunting *shachi*, taking about 1,100 orcas between 1954 and 1997.

In Siberia, the indigenous Yupik people mythologized orcas in a fashion similar to that of Northwest Coastal tribes, as wolves who transformed into sea creatures. The Yupik believed that orcas helped them hunt walrus in the summertime and then became wolves that seemed to assist them in rounding up reindeer in the wintertime. In any event, the Yupik clearly believed they had a mutualistic hunting relationship with their fellow apex predators.

Western mythology casts killer whales in a much darker light, however, as made clear by their given name *Orcinus orca*—from the

Latin "Orcus," one of the Roman gods of the underworld—the one who punished broken oaths. He became associated with demons and was the source of J.R.R. Tolkien's word for a kind of goblin: orcs. *Orcinus orca* roughly translates as "demon from hell."

That stands in stark contrast to their fellow dolphins (particularly the Atlantic bottlenose), who figure prominently as friends to humanity throughout Greek and Roman mythology. Where dolphins were seen as messengers of the sun god, Apollo, orcas in contrast were the minions of Hades.

Indeed, orcas have pretty much always scared the hell out of Western Civilization. They frightened Pliny the Elder, who provided the first written description of them in 70 CE: "Orcas (the appearance of which no image can express, other than an enormous mass of savage flesh with teeth) are the enemy of [other whales] . . . they charge and pierce them like warships ramming." The better part of two millennia

Seen in the Louvre, Jean-August Dominique Ingres' Roger Delivrant Angelique *features a fanciful version of a seagoing "orc."*

later, they scared the bejesus out of the whaler Captain Charles Scammon, who described them in 1874 as "intent upon seeking something to destroy and devour."

For the most part, they were feared and shunned on the seas by most seagoing men from the West for the better part of two millennia, and mythologized by those who remained on land. A fairly typical example of the conception of the "orc," the sea monster from the deep, can be found hanging in the Louvre in Jean Auguste Dominique Ingres' 1819 painting *Roger Delivrant Angelique*, which depicts a hero from Ludovico Ariosto's epic Italian poem, *Orlando Furioso*, astride a winged hippogriff, rescuing a radiant and naked maiden, who is chained to a rock amid a dark roiling sea, from the black monster. The monster's head and face more resemble the crude (and similarly off) medieval depictions of lions, being rather doglike with a mouth full of sharp teeth into which the hero is driving his lance. Gustave Dore's earlier etching of the same scene showed a much more dragon-like creature with wings and a sinuous body.

According to Ariosto, the orc haunted the waters around Ireland and devoured both men and women, though it apparently developed a taste for beautiful women that was never sated. Ariosto described "the monstrous orc" thus:

> *What this resembled best,*
> *But a huge, writhing mass, I do not know;*
> *Which wore no form of animal exprest,*
> *Save in the head, with eyes and teeth of sow.*

Roger, in fact, is unable to defeat the creature because it is too powerful for him, but he manages to stun it briefly with a magic shield and escape with the maiden. (It's also perhaps worth noting that, in Ariosto's poem, the maddened hero himself promptly attempts to sexually assault Angelica after dispatching the orc, but fails because he can't remove his armor.)

Spanish-speaking Basque whalers, who pursued some of the same baleen whales as orcas and saw them as competitors, named them *asesina baleenas*, "whale killers." The name, transposed and slightly mistranslated in English as "killer whales," stuck.

Beginning as early as the 11th century and peaking in the 16th century, these same Basques were the first Europeans to hunt and harvest whales commercially. Mostly they appear to have harvested Atlantic right whales and gray whales, but by the early 18th century, their whaling techniques—they killed both mothers and their calves—led to a serious decline in the cetaceans' population, and the industry withered to nothing.

Of course, by then, other Western nations had joined the hunt, and empowered by sailing technology that made it possible to ply craft across all the world's seas, it became a global one. In the 18th century, many large whaling operations, primarily in America, and much of Europe, began roaming the world's oceans in pursuit of great whales, particularly the mighty sperm whale, whose huge echolocation melon (the "casing" prized by whalers) produced the finest machine oil known to man and whose processed blubber helped light millions of whale-oil lamps, thus fueling the Industrial Revolution. Killer whales were too small to be of interest to these hunters, although many sailors observed them dining on the whalers' catches that were pulled up to the ships for the process of blubber removal known as flensing and many feared the consequences if those rows of flashing teeth should come for them, which, by most accounts, they rarely if ever did. They were also observed chasing away sharks from the kill, which made sailors wonder if they were perhaps to be welcomed instead.

There was one whaling operation, however, that not only interacted with killer whales, but had a symbiotic relationship with the orcas in the manner of the Yupik. They called these orcas "the killer whales of Eden." Eden is a small township near the coast of Australia's New South Wales, near where a Scottish family named Davidson had set up a homestead at the mouth of the Kiah River—two homes, actually, one on each side of the river. About 1840, the family began hunting the Southern right whales who came up to the river's mouth to feed, going out in kelly-green rowboats to harpoon and eventually kill the creatures and then rendering their blubber for oil. They soon found that they had partners in whaling in a large pod of orcas who were also hunting baleen whales. After observing how the orcas helped them, the Davidsons figured out a working arrangement.

The orcas would drive the whales into the river's mouth, alerting the humans to the presence of prey by extraordinary displays of tail-lobbing and wait for the humans to harpoon and kill them. Then the Davidsons simply tied a rope and buoy around the whale's flukes and left the first spoils to their cetacean partners, who would feed on the carcass for a couple of days, mostly taking just the lips and tongue (just as they usually did when humans were uninvolved). The whalers then would row out and bring the remains in to shore for rendering. They called this arrangement "the Law of the Tongue."

This went on for 90 years. Over time, the Davidsons and the orcas developed a special kinship. If orcas became entangled in the whalers' ropes, the humans would abandon whatever efforts they were engaged in to rescue their partners. These were also shark-infested waters, and occasionally the big baleen whales would smash the little green rowboats and send their occupants spilling into the sea. When this happened, the orcas would surround the men in the water and protect them from sharks until they could be rescued.

The orcas were selective. Other whaling operations would periodically show up in Eden to partake of the bounty there, and the orcas would simply ignore them. When it came to the Davidsons' trademark-green boats, however, the orcas were so helpful that they would even grab ropes in their mouths and assist with the hauling.

At the time, these orcas and their exploits were a media sensation. Numerous newspaper stories told their saga, and members of the Australian Parliament came out to Eden to watch, along with hundreds of other spectators. One of the first documentaries ever committed to film, a 35-mm feature shot in 1910, recorded the killer whales in action. All prints of the film, other than a few stills, were reportedly lost in a vault fire in 1916.

The orcas also became well known individually and were given names such as Hooky, Humpy, Cooper, Big Ben, and Big Jack, although none was more famous than Old Tom, a massive male who was noted for hauling with ropes. When Tom died in 1930 at the age of 35—his body washed up on shore, and his skeleton still hangs in the local museum—it largely marked the end of the era; by then, few baleen whales were showing up, and the whaling operations soon went out of business. The memory of beneficent orcas, too, faded quickly.

Indeed, aside from the Eden episode, the Western record of interaction with orcas is largely one riddled with superstitious fear, more a product of the awe that is the natural human reaction to being in the presence of these creatures than a result of any actual attacks. What most impressed early observers of killer whales was their efficiency as predators.

"Three or four of these voracious animals do not hesitate to grapple with the largest baleen whales, and it is surprising to see those leviathans of the deep so completely paralyzed by the presence of their natural, although diminutive, enemies," wrote whaling captain Charles Scammon in his 1874 book, *The Marine Mammals of the Northwestern Coast of North America*. "Frequently the terrified animal—comparatively of enormous size and superior strength—evinces no effort to escape, but lies in a helpless condition, or makes but little resistance to the assaults of its merciless destroyers. The attack of these wolves of the ocean upon their gigantic prey may be likened, in some respects, to a pack of hounds holding the stricken deer at bay."

This view of orcas as merciless destroyers remained intact for most of the next century. Fishermen in the Northwest frequently brought rifles along in their boats for the specific purpose of shooting at killer whales should they encounter them, both out of fear and out of a belief the "voracious" orcas were competing with them for salmon. When they showed up near coastal communities, mothers would gather their children and put away their boats and life jackets to ensure they did not go out on the water when the fearsome creatures were present.

Even the people who would eventually lead the way in changing perceptions about orcas held this view. One of these is John Ford, who eventually grew up to become one of the world's leading killer-whale scientists. "I had a sort of memorable encounter with my family in our little sports fishing boat out of Sooke, just west of Victoria, out salmon fishing and a big pod went by, heading out of Juan de Fuca," Ford recalls. "At the time—this would have been the early '60s—of course, we were all terrified, because they had such a fearsome reputation in those days that they could upset us and we could be ripped apart. I do have a vision etched in my visual cortex memory of the white markings of the whale as it passed under the boat."

The renowned undersea naturalist Jacques Cousteau, musing on an encounter with orcas in 1967, observed that "we, along with the rest of the world, considered killer whales to be the most fearsome creatures of the sea, the avowed enemies of all life forms to be found in the water, including divers."

On what was this reputation founded? If you examine the historical record, almost nothing. Whalers at work processing slain hunted whales at shipside sometimes claimed that orcas would try to grab them, but that was most likely a case of mistaken identity or of simply coming between the orcas and the meat they sought. In 1910, explorers with Robert Scott's doomed Terra Nova Expedition in Antarctica reported that killer whales attempted to tip an ice floe on which a photographer and his dogs were standing, although it was likely the similarity of the dogs' barking to the seals' barking that attracted the orcas.

Likewise, the only recorded attack on a human being by a killer whale involved another likely case of mistaken identity. In 1972, a surfer named Hans Krestchmer at Point Sur, California, was drifting on his board one day, looking rather like a seal, when an orca grabbed him from below by one leg and dragged him down but quickly released him after Kretschmer hit it with his fist.

Earlier that same summer, a pod of orcas attacked the *Lucette*, a family sailing vessel near the Galapagos Islands by ramming it; the 43-foot yacht quickly sank, but the orcas did not attack the people aboard who scrambled into an inflatable raft. The family was forced to endure a 37-day survival odyssey northward before being rescued.

There was, as it happens, a long-ago antecedent for the attack on the *Lucette*. The survivors of the sinking of the whaleship *Essex*, which had been rammed by an angry sperm whale in the South Pacific, the maritime disaster that inspired *Moby Dick*, reported a similar attack on one of their lifeboats during their ordeal, which lasted a full 89 days at sea and left only eight survivors out of twenty men. This orca had appeared out of nowhere and taken a bite out of the side of the lifeboat, then rammed it and split the boat's stem. It was driven off by men with poles, but it had shaken these whale hunters and made them feel as if they had become prey. "We were not without our fears that the fish might renew his attack, some time during the night, upon one of the other boats, and unexpectedly destroy us," the captain later recalled. It did not, however, and the incident remained a singular one until 1972.

The father of the family that survived the *Lucette* sinking later speculated that his hull might have resembled a humpback whale. However, he also assiduously warned of the dangers posed by killer whales, should this kind of attack be repeated or develop into a trend. As it happens, it has not, to anyone's knowledge.

A 1963 book by a science popularizer named Joseph J. Cook (who also penned tomes on sharks and Atlantic herring) titled *Killer Whale!* represented itself as an objective, scientific look at orcas and their "vicious instincts," which, according to the authors make them "the fiercest, most terrifying animal in all the world," a creature "capable of attacking anything that swims, no matter how large. They are afraid of nothing, not even boats or ships."

"The killer whale is well designed for a career of destruction and mayhem," Cook and his coauthor, William J. Wisner, wrote. They contrasted orcas with their cousins, the bottlenose dolphin. "How different the orca, which seems to be filled with a burning hatred! Nothing that lives or moves in or on the water is safe from its assaults. Its size, power, speed, agility and disposition have made this black monster feared wherever it is known."

Then someone got the clever idea to capture one of these terrifying creatures and to put it on display. And that changed everything.

• • •

It was the go-go Sixties, and dolphinariums were flourishing. Following the success of a Florida park called Marine Studios, which had discovered in the late 1930s that dolphins could be trained to perform stunts and which began putting on dolphin entertainment shows that packed in large crowds, aquarium owners around the country got into the business. By the 1950s, they were everywhere, with larger and larger facilities selling more and more tickets and taking larger numbers of dolphins into captivity.

One of these was Marineland of the Pacific, an oceanarium owned by Marine Studios near Los Angeles, located right on the Pacific Ocean at the Palos Verdes Peninsula. When it opened in 1954, it was the largest such facility in the world and was one of California's biggest tourist draws (after Disneyland) for years. Initially, it specialized in truly acrobatic displays featuring Pacific white-sided and spinner dolphins.

Then, in November of 1961, a single, disoriented female orca wandered into Newport Harbor, several miles south of Marineland, and harbor officials contacted Marineland about the animal. Their dolphin-capture specialists traveled to the harbor and tried to get control of the whale using their usual techniques and failed, but eventually they managed to net her, place her on a flatbed truck, and trundle her the fifty miles or so to Palos Verdes. Once in her concrete tank, however, the female (whom they named Wanda) went berserk. "We'd suspected the animal was in trouble because of its erratic behavior in the harbor," recalled Frank Brocato, Marineland's chief animal captor. "But the next day, she went crazy. She started swimming at high speed around the tank, striking her body repeatedly. Finally, she convulsed and died."

The necropsy placed the blame on pneumonia and acute gastroenteritis, but the experience gave Brocato and his bosses an idea: Why not see if they could keep one of these huge dolphins captive for a longer duration? The following September of 1962, he and his crew arrived in the Salish Sea with their 40-foot capture boat, the *Geronimo*. Eventually, off the west side of San Juan Island on a foggy day with gray mist hovering over the water, they managed to find a couple of orcas who were in pursuit of a harbor porpoise. The porpoise approached their boat, hiding around its hull, in the hope it could use the craft to evade the orcas. Brocato realized he could reach a female orca as it pursued the porpoise around his boat, so he got out a large lasso and nabbed her as she went past.

"But then everything began to go wrong," recalled Brocato. The female twisted and dove under the boat, pulling the heavy nylon rope right into the propellers and killing the engine. Then she ran to the end of the 250 feet of her tether and surfaced, vocalizing loudly with fearful cries. At that, the large male accompanying her, appeared out of the mist and charged the *Geronimo*. To Brocato and his crew, it appeared that they were about to be attacked, but the orca dove at the last moment and whacked the hull with its flukes. This happened several times. Brocato ran and got his high-powered rifle and began shooting.

He managed to shoot the male once, after which it dove and disappeared. The female continued to struggle, and Brocato, now fearful for his craft, began shooting at her. It took ten rounds to finish her off. He towed the body into Bellingham, Washington, where it was rendered for dog food. Marineland decided that capturing orcas was not going to work out; they seemed too wild.

Finally, in 1964, an artist commissioned by the government in British Columbia to kill an orca in order to create a life-size statue, managed to harpoon a young male near Saturna Island, just across the Canadian boundary from the San Juan Islands. The whale was only wounded by the harpoon, which remained embedded in his flesh (the hunters also tried shooting the whale, to little apparent effect); soon, the manager of the Vancouver Aquarium arrived and took possession of the orca, having decided to try keeping it alive. Presumed to be female because of his fin size, the male whale was dubbed "Moby Doll" by the local press, which descended on the scene, along with thousands of spectators, as the whale was towed into a makeshift pen in Vancouver harbor.

Moby Doll refused to eat for his first fifty-five days in captivity, during which time he appeared to be in shock, floating listlessly in the pen. Then, without warning, he began eating fish again, voraciously so, and accordingly became much more active. However, he also developed skin lesions because of the harbor's low salinity, and although he ate well, he did not look healthy. Sure enough, some thirty days after he had resumed eating, Moby Doll died, at which time a necropsy revealed him to be male. He had lived 87 days in captivity.

The killer whale created such a media sensation, however, that now the demand from aquarium owners became real. When two fishermen using purse-seine gillnets near the village of Namu, about 100 miles north of Vancouver Island along the British Columbia coast, accidentally netted a large male orca, they decided not to let it go, as they normally might. They had heard the stories of big money for such a catch.

The men contacted Ted Griffin, proprietor of the Seattle Aquarium, who flew up the next day and paid $8,000 in cold hard cash for the whale, which he then towed back the four hundred miles to Seattle in a floating cage. The cage attracted a pod of orcas who then accompanied the captive whale much of the way back to Seattle. Plaintive cries could be heard from the caged orca, although his pod gradually drifted away as the cage neared Puget Sound.

Griffin dubbed the whale Namu after its home village and had an immediate sensation on his hands. Tens of thousands lined up to see the five-ton killer whale, which he kept in a sea pen at the aquarium on Elliott Bay along the Seattle waterfront. Griffin came to realize that the big male was a benign creature and began climbing into the pen to swim with him, giving the whale back scratches and developing a bond.

A Seattle rock band dubbed themselves The Dorsals and put out a 45 rpm record titled "Killer Whale": "Gotta killer whale in tow, Namu in the sea! Gotta killer whale in tow, glad he's not after me! Five tons of whale in tow, and that's a great big thing. Movin' in his playpen, Namu's really king."

The footage of Namu being lowered by a crane and sling into his pen at the Seattle Aquarium, with traffic whizzing past behind him on the Alaskan Way viaduct, caught everyone's attention, and Griffin was proved to be cannily accurate about how many tickets his killer whale would sell. Better yet, he soon discovered that not only was the killer whale able to survive captivity in the waterfront pen (given a daily diet of 400 pounds of fresh salmon), but Griffin himself found he could form a personal relationship with the creature. Griffin started out swimming with the whale and then, sometime later, first his brother Jim and then Griffin clambered aboard its back and rode around the pen, hanging onto the orca's big dorsal.

It was enough to make him a movie star. And it did. This is how it happened.

• • •

Ivan Tors liked to make a certain kind of nature show. They were all, movies and TV series alike, scripted family dramas. Mostly they featured wild animals who, defying expectations (sometimes they were huge, potentially dangerous beasts such as the "crosseyed" lion Clarence in his TV series *Daktari,* or the large black bear who was pals with a boy in *Gentle Ben*) befriended humans, invariably a child and his or her adult guardian.

Tors, originally a science-fiction movie producer (ever hear of *Gog?*) who specialized in featuring a mix of real science in his stories, hit upon this reliably moneymaking family-friendly formula when he produced the 1963 movie *Flipper*. Tors' productions had taken an aquatic turn after they had a hit with the 1958-'61 series *Sea Hunt,* which launched the career of Lloyd Bridges. The story of a dolphin befriending a 12-year-old boy and winning over his fisherman dad (Chuck Connors) was a major box-office success if not an outright smash.

The next year, *Flipper* became a hit television series starring Brian Kelly as the gruff father, and it remains oddly influential and durable. Indeed, it's hard to find a marine biologist of a certain age who does not credit *Flipper* with creating his or her interest in the field. It was revived in the 1990s as both a movie and a TV series.

Tors was in the middle of his run with the *Flipper* series when the stories about Ted Griffin and his killer whale started making national headlines. Tors himself showed up at the Seattle Aquarium to talk to Griffin about making a movie starring Namu. He told Griffin and his brother, Jim, that they could only make such a film if someone could actually ride the whale. That led to the first attempt to do so.

Griffin recalled it years later for PBS's *Frontline*: "And so while I was thinking about it, my brother went to the back room, put on a wet suit, climbed in the water with the whale and went swimming with the whale. So in point of fact, he was the very first. I hate to say that because while I was thinking about it and enjoying this period of time, Jim, practical man that he is, decided that the contract was extremely important and they had to have footage of a man and a killer whale together. Well, they had their footage in an hour, and Jim was out of the water. And he said, 'I got all my arms and legs. Go ahead. The water's safe, you know.'"

Not only did the stunt persuade Tors, it also persuaded Ted Griffin to try it himself. Soon the whale riding became a regular part of his frequent interaction with the whale, but it went beyond that. Griffin had a great deal of tactile contact with Namu, who loved being rubbed. Recalling how the orca had seemed to signal a hello the first time they met in that purse-seine net, Griffin tried communicating with him with surprising results.

"At [the] time, I could squeak to the whale, and the whale would squeak back," Griffin recalled. "I didn't know what I was saying to the whale, but I knew I was learning to communicate something and getting the whale to respond."

Griffin invited a Stanford scientist named Thomas Poulter to record the sounds in order to figure out if something real was occurring. "He had notes about who said what and when it was said. … And he said, 'You're not talking like the whale. The whale has entirely changed his vocalization to sound like you. And here it is on the sonogram.' And I was just thunderstruck. But probably one of the biggest shocks that I ever had was that this whale was trying to reach me in this way."

When it came time to make the movie, Griffin towed Namu out to a bay near Port Orchard where it could appear as though Namu was in the wilds of the Northwest. Griffin performed most of the stunts that are seen in the film, although the star, Robert Lansing, also spent time in the water with the orca and filmed several scenes riding him as well and coaxed Namu into performing a number of breaches and other maneuvers.

The movie that resulted, titled *Namu, the Killer Whale* (later retitled, for video release, *Namu, My Best Friend*), was an archetype of the Ivan Tors genre, although much less successful than many of his projects. It was heart-warming family fare with a slightly provocative thesis: What if killer whales were not vicious killers at all, but instead rather gentle, intelligent creatures capable of friendship with human beings? As with nearly everything produced about orcas in this period, it's pretty amusing to watch *Namu* now, in light of what we have learned about the species in the years since. For all that, it encapsulates an important moment in the history of human-animal relations, a tiny moment signifying a sea change, when people first began to realize that orcas were not only benign but also intelligent.

Like most of Tors' adult male heroes, Robert Lansing was a square-jawed, athletic, clean-living type of male lead. His character, Hank Donner, is a marine biologist who has set up a field camp on the shores of a remote spot in the San Juan Islands. He erects a sign onshore announcing that it's a "Marine Research Station" and builds a dock so he can observe underwater wildlife. One day Donner witnesses a boatful of local yokels out taking potshots at killer whales from their boat. One of them, a rangy sort with a mean streak played by John Anderson, scores, severely wounding a female, who comes into shore at Donner's camp and dies there. She is accompanied by her mate, a large male played by none other than Namu, who then remains inside the cove, mourning her death.

For people watching these movies for laughs, this setup provides a few chortles. Orcas, we now know, never remain with their mates, and they likely never mourn them or, as Namu does in the movie, "call for his dead mate." They remain with their mothers their whole lives (although if the script were rewritten to make this female the male orca's mother, that would in fact be perfectly credible). Also, it doesn't help that the sounds they give to Namu, both above and below water, sound little like actual orca calls, but more like chimp noises grafted on by the sound-editing crew.

After Donner chases the hunters away from his outdoor lab, the orca inexplicably remains there, so Hank goes to town to buy a net to keep him in "so I can do some research on him" and to keep him safe from the locals. This is where little Lisa Rand (played by 10-year-old Robin Mattson) first hears about the orca, and soon she and her school pals are down at the cove, learning all about the whale from firsthand experience and watching Hank work with him. Hank, of course, eventually rides Namu, to the astonishment of all, especially Lisa's mom Kate (Lee Meriwether, in the days before she became Catwoman and Barnaby Jones' secretary).

Then one of the yokels' kids pulls a stunt that puts a mouthful of hooks in Namu's mouth, and he reacts angrily, scaring the kids away and sending their parents marching down to the cove with rifles (or is that torches and pitchforks?) to string that whale up before it eats their kids. Hank stands on the dock with his friends behind him and prevents the lynching, and in the process releases Namu from his sea pen. However, Mr. Mean Streak follows him out with his rifle, tries one last time to shoot the orca and is thrown into the water with the killer whale. Of course, he can't swim, and Namu saves him. Everyone now loves Namu. The End.

· · ·

It probably didn't help the box office for *Namu, the Killer Whale* that its hero had died only two weeks before the movie was released. It quickly sank out of sight.

Ted Griffin always blamed Namu's death on the water of Elliott Bay at Pier 56, where the orca was kept in captivity; Griffin believed the pollution made the orca sick. Whatever the cause, the whale contracted a bacterial infection, and overnight became entangled in the cable nets at the aquarium and drowned. Griffin had to go retrieve his body.

"It was after dark and I had to go down in the water and I had to cut him out of the nets," he recalled. "He was in an enclosure. And it was a very difficult time. You pull yourself together and you think OK, this has to be done and there's nobody else to do it.

"There are those brief moments when the light from the docks reflect in his eye and you think you see movement and you allow yourself to believe the whale's still alive. I did. [But] I toughened up and held together until I was able to get the whale free of the enclosure and to another area where he could be removed from the water. And at that point, I was a total disaster and had to leave the area. Fortunately my partner, Don Goldsberry, was available and he being somewhat tougher than I am in this respect was able to take charge and manage the affair."

Griffin was emotionally devastated—he took up poetry, and drifted away from the aquarium work—but not enough to reconsider whether captivity was right for the whale he had befriended or any future whales, either. "I was resolved to have another pet, so I continued capturing killer whales, and a year later I did acquire additional whales," he recalled. At first he started to make plans for another orca in Seattle, and then realized that he couldn't bear it.

"Couldn't do it. And so at that point, it was pretty much over for me," he said. "So I pulled things back together and started acting like a businessman, and we captured whales for zoos and aquariums around the world. Japan, Germany, England, France, and of course SeaWorld and others."

Griffin and his partner, Don Goldsberry, had perfected the technique that began with that first successful capture. Even before Namu's death, with an eye toward the future, they had captured a young female. That capture had not turned out so well. The young orca had been alone with its mother when Griffin and Goldsberry had cornered them in a cove near Tacoma, Washington called Carr Inlet. Griffin harpooned the mother, intending her to be a female companion for Namu. However, the wound was deep and mortal; the captors towed her back to Seattle, and her young calf followed. Eventually the mother died, drowning herself, but the calf had been netted by then, and Griffin decided to keep her.

When they put the young calf in the same pen with Namu, the traumatized young whale responded enthusiastically to the big male, too much so. She rammed him frequently. It caused a problem for her interactions with Griffin especially: "Interestingly enough, the whale began—maybe because she was young—to roughhouse with me in a very dangerous way," he recalled. "And when I [was] in the water with Namu, even riding him, this whale would ram Namu with such force that it might have killed me. And when I was in the water with Namu . . . the whale would ram me, but not as [seriously] as some of the times that she rammed Namu."

At the time, he had representatives from Sea World, a big new facility in San Diego, visiting. They were very interested in acquiring a killer whale, and so he offered them the female. They agreed and wanted to buy the rights to the name Namu along with it. Griffin refused. Instead, they agreed to a concocted name for a female Namu, combining "She" and "Namu" to make "Shamu." (SeaWorld guides at one time told customers the name means "Friend of Namu" in Salishan, but this is a fabrication.) And while there would be no more killer whales at the Seattle Aquarium, there would be many more Shamus.

The original Shamu lived another six years at Sea World San Diego. She was always somewhat irascible, but generally had a solid reputation as a performer—at least, until that day in April of 1971 when, as a publicity stunt, Sea World publicists put a hapless front-office secretary named Annette Eckis, wearing a bikini, into a little tank with Shamu and told her to take a ride. Shamu reacted violently, grabbing Eckis by the legs and biting them, then dragging the shrieking woman around the pool before finally releasing her.

Apparently Shamu had been acting up with her trainers, so it was decided to move her into their breeding program, where there would not be as many opportunities for dangerous interactions. Sea World vets put her on a program of progesterone to increase her fertility, but instead she wound up contracting pyometra, a condition that causes serious infections in the uterus. She died of septicemia from such an infection on August 29, 1971, at about nine years of age. In the wild, her grandmother lived to be a hundred.

However, Sea World gave Shamu a kind of immortality, although it did not release the news of the orca's death immediately. By then, her temperamental nature had forced Sea World to use other whales in her place while calling them all Shamu. It was an easy step to give all their performers the stage name Shamu, for the sake of simplifying things for the audience. Shamu is also the name given to all the cuddly killer-whale stuffed animals they sell in their gift shops, the name on the all the T-shirts and other gewgaws they sell, the name of the stadiums and other facilities in both San Diego and Orlando where killer whales are kept and perform, the very embodiment of the SeaWorld brand name. Given the sordid history behind it, though, it might be more apt to drop the final "u" from her name.

• • •

Quick, name the movie in which Bo Derek gets her leg bitten off.

That's right: *Orca: The Killer Whale*. Released in 1977. Starring that great chewer of lovely scenery, Richard Harris.

It was a quickie ride-the-wave exploitation flick from Dino de Laurentis meant to cash in on the *Jaws* phenomenon. (Reportedly, de Laurentis called up one of his chief producers in the middle of the night after seeing the 1975 box-office champ and asked him to find an animal that was bigger and badder than a shark.) Indeed, the first victim of the titular killer whale in the film is not a human, but a great white shark rather similar to the one dispatched by Roy Scheider and Co. two years previously. So, verisimilitude was not exactly its first mission. It failed at making money, too. It does, however, give a glimpse into the popular mindset about killer whales at the time, as well as the state of some of the pseudoscience that was still being bandied about as settled wisdom about whales in general and orcas in particular. It is kind of a revealing relic of how those attitudes were in transformation.

It also has the virtue of being a spectacularly, laughably bad movie, with all the requisite components: slumming work from great actors (Harris and Charlotte Rampling), early work by a future starlet (this was the only film Derek had made prior to her short-lived star turn as the "it girl" object of Dudley Moore's obsession in 1979's *10*), a pretentious script. Everything in *Orca* is bad: bad writing, bad science, bad acting by everyone involved, and spectacularly bad special effects. Most shots of live killer whales were clearly taken of captive orcas inside a concrete tank; you can see the flopped-over fin, typical for captives, on the male—actually Nepo, a big male captured in Pender Harbor in 1969—in footage taken at the old Marine World park in Los Angeles, while the shots of the rampaging killer whale of the title circling the boat all show a normal straight fin with a bloody notch where Harris' harpoon wounded him. Of course, that circling fin was quite obviously fake as well.

There's the bad science, all of which has been superseded since 1977 by actual research into killer whales and their familial structures. "You know, killer whales are monogamous," warns Bo Derek, fearful that Harris's attempts to snare a killer whale might "break up a family." Perhaps this is why the angered male orca later bites off one of her legs. Really, it tells you everything you need to know about the film's relationship to scientific reality that, when the titular orca sees its mate and unborn child being killed, it raises its head about the water and roars, loudly, deeply, like a lion, with rage.

"Like humans, orcas have a profound instinct for vengeance," relates Rampling, who plays the film's resident killer-whale researcher. (She also plays recordings of humpback songs to explain to her students how orcas communicate.) This evidently explains why, after seeing its family slain by Harris, the orca proceeds to blow up much of the local town (it manages to rupture a fuel line and set it afire, which spreads to a refinery) and then tears down the shoreside building housing Ms. Derek, who slides with the wreckage into the water and loses her leg to the voracious orca, courtesy of the aforementioned special-effects realism. (I think Derek's missing leg turned up a few years later as the lamp in *A Christmas Story*.)

The film trundles on to a weird climax aboard Harris's boat in the Arctic waste (actually filmed in the warm waters of Malta) during which Will Sampson, desperately looking for something to do after *One Flew Over the Cuckoo's Nest*, gets crushed by fake falling ice that sinks their boat. Then Richard Harris confronts his tormentor on a fake ice floe, slips into the drink, and is flipped fatally out of the water by the orca's powerful flukes back onto the ice, whence his body then slides down back into the water and sinks.

That's right. Unlike *Jaws*, this movie features a creature who wins and is still alive at the end. No doubt Dino de Laurentis noticed that killing off the shark at the end of *Jaws* made a sequel much trickier, so he dispensed with that problem here, and if *Orca* had not bombed, no doubt there would have been a sequel. Reportedly, de Laurentis toyed with the idea of having the monster gorilla he created for his 1976 remake of *King Kong* face off against the orca in a 1980s movie-monster sequel—sort of like *Godzilla vs. Megalon*, without the wrestling moves—but couldn't get anyone to bite.

Convenience aside, the film's outcome is striking in its rarity. Having the monster win was unheard of in these man-vs.-beast epics. Even more striking was that it was the most satisfying outcome, given that Hollywood is built on revenge melodramas such as this, but with humans usually getting the revenge for harm inflicted by the titular animals. In this movie, that's all turned on its head; the audience is far more inclined to sympathize with the orca, horribly wronged by the Harris character, who despite displaying some pangs of conscience nonetheless is a blithering idiot from start to finish and a despicable brute in the early goings.

In this regard, it was one of the first films to portray an animal as not merely intelligent but also as a moral actor, which reflected, similarly, the public's shifting perceptions of killer whales generally. This would culminate, a few years later, in the movie *Free Willy*, the public campaign to "rescue" the whale made famous in that film, and the growing agitation to release captive killer whales back into wild waters.

It must be said, however, that movies had only a secondary role in the initial transformation of public attitudes about killer whales; if anything, they only reflected changes that came bubbling up along with the rise of an environmental ethos, especially in the West, embodied in the various "Save the Whales" campaigns that arose in the late 1960s and '70s.

Make no mistake: The sea change in public perceptions about killer whales that occurred then was primarily a product of aquariums and marine parks that put captive orcas on display. Killer whales ceased being demonic bringers of death, all teeth and vengeance, and almost magically became transformed instead into cuddly watergoing pandas, friendly to humans, endlessly docile, and pretty darned smart, too. It wasn't quite as false as the old narrative—but it came close.

• • •

Ted Griffin and Don Goldsberry kept taking killer whales out of Puget Sound waters, and a number of other captors operated similarly in the Vancouver Island area, up until the mid-1970s. By then, everybody had gotten a good look at how they went about their business. It wasn't pretty. In addition to setting purse seine nets to snag the whales, Goldsberry and Co. took to the skies to herd the whales into those nets, dropping seal bombs, small explosive devices, from the planes to frighten the orcas into their well-laid traps. It may have been brutal, but it worked. Of course, the brutality of it all meant that inevitably there were some collateral losses, as the orca cowboys might have put it. A lot of whales were dying in the process.

One of the worst of these occurred in August 1970, when Griffin and Goldsberry and crew drove a large pod of whales into a large bay on the eastern side of Whidbey Island called Penn Cove, renowned for its mussels and clams, within view of most of the residents of Coupeville, one of the largest towns on the island. It also occurred within earshot, and the noise was terrible; to this day, people who witnessed the roundup talk about how the whales "screamed" when they came up and spyhopped. A large group of other whales gathered outside the net and exchanged whistles and calls with those inside.

In the process of collecting seven whales for distribution to marine parks around the world, the crew also managed to kill four orcas, three calves and an adult, who drowned in their nets. Ted Griffin ordered workers to surreptitiously slit the whales' bellies, weigh them down with chains and tires, and sink them. Unfortunately for Griffin, several months later, the whales' corpses were caught in a fisherman's net and pulled up onto shore, where news photographers recorded the event, and considerable outrage grew about the tactics being used to capture these whales.

Because they were such a common sight for people who plied the waters of Puget Sound, the assumption was that there were thousands of these whales and that removing a few for captive display wouldn't harm the population, but then scientists began studying them more carefully and discovered that it was closer to a hundred whales all told—if that many.

The study techniques were pioneered by the Canadian marine biologist, Michael Bigg, who figured out that the white "saddle patches" behind a killer whale's dorsal fin are unique to each animal, like a fingerprint. Moreover, the dorsal fins were often distinctive in themselves (particularly those of males and those that had nicks in them) and could offer quick identification of individual whales. By photographing them, Bigg and his fellow scientists were able to compile an accurate census of the killer whale populations in the Pacific Northwest. By the mid-1970s, the shocking news was confirmed: There were fewer than a hundred whales in the population of orcas that lived in Puget Sound and the San Juan Islands, and there were about two hundred whales who lived in the area of northern Vancouver Island and the archipelago of fjords and islands around it.

The issue came to a head in March 1976, just as the Washington Legislature began debating whether to call a halt to any further captures of killer whales from their waters. It happened that Don Goldsberry and his crew, now working for SeaWorld, drove a pod of whales into the harbor in Olympia, in view of the capitol dome where the matter was being discussed, as well as within a stone's throw of the college campus where a conference of whale researchers was getting under way.

Goldsberry was using his usual techniques to harass the whales, dropping seal bombs, and buzzing the whales with airplanes. It was obvious to everyone watching that the roundup was in violation of his permit under the recently passed Marine Mammal Protection Act of 1972, which required collectors to be "humane." Washington state officials, including Secretary of State Ralph Munro and Attorney General Slade Gorton, took SeaWorld to court and shut down their operations in Washington State permanently. The legislature, meanwhile, declared Puget Sound to be a sanctuary for the whales. Canada did likewise for its waters, shutting down orca captures in the Northwest for good.

The whales of Puget Sound do not often return to the places where they were captured. With rare exceptions, you will no longer see killer whales, as you once did, in Penn Cove, Carr Inlet, Port Madison, Yukon Harbor, or any of the places where humans drove them into nets and captured them. They still visit many diverse places in the Puget

Sound's waters, are still seen from time to time chugging right past Seattle itself, but they do not go to those other places anymore. They seem to remember.

• • •

Every one of the fifty Southern Resident whales captured by Ted Griffin and Don Goldsberry from the Puget Sound is now dead, with one exception. Lolita, one of the young calves taken at Penn Cove in 1970, remains alive 44 years later at the Miami Seaquarium. Griffin lost the stomach for orca captures after the Penn Cove debacle and dropped out after 1972. Griffin said he could see the writing on the wall, probably in the form of the Marine Mammal Protection Act, which was passed that year and which placed tough conditions on capturing the creatures.

Goldsberry packed up his operations and went north, far north, to Iceland, where he and his colleagues knew there was another orca population ripe for the plucking, especially since Icelandic fishermen had always seen orcas as competitors for herring. Working for SeaWorld, they went to work and removed some 48 whales from the orca population around Iceland, between 1976 and 1988, and they had plenty of customers.

The orca-display industry blossomed in the 1970s and '80s, especially as aquarium owners realized they could draw large crowds with the big mammals. They sprouted in places like Ontario, Hong Kong, San Antonio, Munich, and the Netherlands. Some of these were former dolphinariums, such as Miami Seaquarium and Reino Aventura in Mexico City, which simply expanded their pools to accommodate killer whales, meaning that the whales were in smallish facilities. Others, such as Sealand in Victoria, British Columbia, were just glorified sea pens constructed mostly to draw in tourists, cashing in quick bucks on the new animal sensation.

The king of the hill was (and still is) SeaWorld, which not only had the largest orca displays in the world in both San Diego and Orlando, but also began consolidating the industry in the 1980s under its corporate banner. Over the years, SeaWorld's ownership has changed several times. The publisher Harcourt Brace Jovanovich bought it in 1976, then sold it to beer-maker Anheuser Busch in 1989; most recently, it became a subsidiary of the Blackstone Group. However, in all those years, SeaWorld has been actively acquisitive, sometimes buying facilities just to take possession of their killer whales and other animals and to ship them to their own tanks, then closing down the aging parks, as they did when they bought Marineland of the Pacific in 1987.

All this was possible because millions of people began filing through the turnstiles at these parks and still do. The marine-park industry attracts more paying customers than even the most popular sports leagues. In 2012, orca facilities around the world drew over 120 million people, more than the combined attendance at Major League Baseball, National Football League, and National Basketball Association games. Orcas are Big Money now.

The public's fascination with orcas was remarkable, considering that less than a generation before, these creatures had mostly elicited shudders of fear. Now the image has shifted almost completely in the opposite direction, to that of giant, friendly oceangoing pandas, big dolphins who are very smart, too. It was true in a way, but a fantasy at the same time.

Nonetheless, it was a brilliant piece of marketing. In addition to tickets, it sold millions of plush orca dolls and orca T-shirts. There is little doubt that, by letting people see for themselves their gentle natures, their playfulness, and their intelligence, the captive-orca industry played an overwhelming role in not only shifting public attitudes about killer whales but also of entrancing millions of young people with the sense of awe that orcas naturally elicit. The industry has been enduringly influential; many younger marine scientists today, who were not born when *Flipper* was on the air, credit visits to SeaWorld or similar marine parks with inspiring them to pursue careers in biology.

Along the way, the facilities for the orcas have improved significantly; they now have much more room at most parks than in the early years, and their handlers have largely figured out their medical and dental routines. The displays of the whales, meanwhile, have become increasingly oriented toward entertainment with loud music, fantastic leaps and acrobatics, and amusing displays of cleverness. SeaWorld's shows in particular, until very recently, featured jaw-dropping interactions between the trainers and the whales, who would pick up their little humans with their flippers and their rostrums and go flying through the air with them.

The remarkable part of all this, especially for the trainers and the people working with the orcas, was not only the creatures' docility and friendliness, but also their intelligence. Indeed, the more people worked with them up close, the more that some of them began to wonder just how appropriate it was to keep them captive. The more they taught them cute animal tricks and tried to make the killer whales behave, the more they wondered just who was really in control. However, that was knowledge only for those who worked behind the scenes, and not many of them remained there if they ever expressed doubts publicly.

• • •

The first effort to return a captive whale to the wild was originally supposed to happen in October of 1982. Haida, a big male J pod orca who had been captured in 1968 by Ted Griffin and his outfit with four other orcas (including Skana), had been purchased by Sealand of the Pacific and had been residing in Sealand's open-water aquarium on the shores of east Victoria for 14 years.

There were many dubious and disreputable outfits involved in orca captivity in the early years, and Sealand was one of them. Despite offering its whales the advantage of fresh seawater, Sealand was a hellish prison for its whales. Once the park was closed for the day, its managers did not give its orcas free roam of its smallish enclosures. Instead, each whale was locked away in an individual "holding module" that was only 25 feet by 30 in size and only 12 feet deep. This meant that for the next 12 hours or so, the whales were required to remain motionless in one place, mostly in darkness, even though these are animals, as we have seen, who do not fall asleep, at least not as we do.

They had tried to mate Haida with a number of females. First came the transient whales they had captured in 1970 near Victoria, the albino, Chimo, and her sister, Nootka. Only Chimo remained in the facility with Haida (Nootka was shipped off to a variety of other aquariums before she died, 20 years later, at Sea World in San Diego), but although she was observed mating with Haida, they never produced offspring before Chimo died two years later. Two more females, dubbed Nootka 2 (a K pod female) and Nootka 3 (a transient), were captured from Northwest waters with an eye toward mating with Haida, but each of those whales died within nine months of capture. Haida's last mate, a rescued orca named Miracle, who in 1977 had been found shot and ill as a young calf and returned to health at Sealand, drowned in January 1982, tangled in the nets that made up Sealand's dubious enclosures, before ever reaching sexual maturity.

After that, Sealand decided it was time to capture more whales. However, the legal and political scene had altered radically, with the system much more carefully regulated since the last time Sealand had overseen orca captures. Organized protests against their efforts to seek a permit to capture more whales resulted in an unusual agreement with the Canadian government: Sealand could capture two more whales if it first released Haida back to the wild.

The release was scheduled to occur in mid-October of 1982. However, on the first day of the month, Haida fell ill with a lung infection, and he died three days later. (A necropsy showed no signs of foul play.) So much for the release of a long-captive orca, as well as for Sealand's hopes to capture more Northwest whales.

Instead, they turned to Don Goldsberry's SeaWorld operations in Iceland and purchased from them three whales captured in 1982, two females, named Nootka 4 and Haida 2, and a young male, named Tilikum.

Even though they had all been captured together, the three whales did not have a good dynamic together, at least not with Tilikum, who was chased around the pool and bullied by the two females, leaving him with rake marks all over his body. He frequently had to be separated from the other two because he was small at the time (ironic, considering his later immense size as an adult) and was not inclined to return the aggression.

His trainer, a young Canadian named Colin Baird, said Tilikum was a pleasure to train. Tilikum—or "Tili," as he was known—was popular and "very easy to work with," Baird recalled. "He was very easygoing, he learned quickly, he learned well, very responsive," he said. "You know, he was probably my favorite of the three."

However, Baird said the tense social dynamic among the three orcas made all of them moody. "They have personalities, for the lack of a better word, individual personalities, and they have good days and bad days just as we do," he said. "There were some days, Tilikum would have a certain look in his eye—then I would just say, 'Nope, not getting in the water with him today.'"

One day, a trainer accidentally got in the water with Tili on one of these days. That is when the orca's bad moods metastasized into a tragedy. The Sealand show of February 20, 1991, had been nothing out of the ordinary. Then, near the end, one of the trainers, a 20-year-old champion swimmer and student at the University of Victoria named Keltie Byrne, slipped and fell into the pool. This was something out of the ordinary for the whales, and as she tried to clamber back out, Tilikum promptly seized Byrne by the leg and dragged her back into the water. The three whales then took turns playing with the trainer like a rag doll. She screamed for help. Other trainers came to her rescue but couldn't separate the whales from her. After a while, Keltie Byrne did not return to the surface. Two hours later, they were able to recover her body.

It was the first time in recorded history that a human being had been attacked and killed by an orca. The story horrified people in the Northwest but did not travel much out of the region. A year later, Sealand of the Pacific shuttered its operations and sold all three of its killer whales to SeaWorld. Nootka and Haida went to San Diego. Tilikum was flown to Orlando, where in time he would grow to become the largest captive orca in the world—and the most notorious.

• • •

In 1999, after eight years at Sea World Orlando, Tilikum killed his second human.

Somehow, a homeless 27-year-old man named Daniel Dukes managed to sneak inside the facility, buying a ticket and then hiding himself inside at closing time. Overnight, he went to Tilikum's holding pool. No one is sure why, but he seems to have gotten into the pool voluntarily with Tili, perhaps so he could experience the magic of swimming with an orca. By then, the orca was not just fully grown, but at 12,000 pounds, he was the world's largest captive killer whale. Tilikum's trainers arrived the next morning to find him parading around the pool with Dukes's naked corpse draped across his back. An autopsy found that Dukes had died of drowning. However, not only had Tilikum stripped him of his clothes, but the man's body was covered in bite marks, contusions, and abrasions, and his genitals had been bitten off. John Jett, Tilikum's longtime trainer, observed years later that no one really ever could figure out what happened, but "the guy definitely jumped in the wrong pool."

Officials at SeaWorld (which by now had shortened its name to a single word for corporate purposes, while still referring to individual parks with separate nouns) shrugged off the incident as a bizarre accident involving a person who violated all their safety procedures. However, eleven years later, Tilikum killed another of his trainers, and this time, there was no shrugging.

It was a normal "Dine With Shamu" show, a program in which people pay large sums to sit at a poolside table and get an up-close look at the whales, at Orlando on February 24, 2010. The show featured Tilikum and one of SeaWorld's top trainers, Dawn Brancheau, an athletic and attractive 40-year-old who had been an orca trainer for 16 years. During the show, there had been a minor misunderstanding between Brancheau and the orca, but no one noticed any tension or problem. At the show's end, as everyone began to file out, Brancheau went to the shallow bench of the "slide out" on the pool's edge and lay in the water, appearing to be conversing with Tilikum. Suddenly, she disappeared under water, yanked into the main pool by the orca. Dawn probably drowned about ten minutes into the assault; Tilikum refused to release his trophy and spent the next half hour battering her corpse further. In the end, when trainers forced him into the medical pool and raised its floor so that they could retrieve Brancheau's body, he madly thrashed about with her body and finally bit off her whole left arm. No one realized this at first; they had to raise the floor a second time so that someone could retrieve it from his mouth.

In the past, SeaWorld had managed to paper over what turned out, on closer examination, to be a long record of assaults on trainers by other killer whales, not just by Tilikum. Now it all came out. The federal government was the first to descend. The Occupational Safety and Health Administration's California body, Cal/OSHA, had investigated Sea World's San Diego facility after an incident in which the female orca Kasatka had seized another veteran trainer, a man named Ken Peters, and had toyed with him about the pool before finally releasing him. That 2007 report had presciently concluded, "If someone hasn't been killed already, it is only a matter of time before it does happen."

Now OSHA's main branch jumped in with both feet and held an investigation into SeaWorld. On August 23, 2010, it announced it was fining the park $75,000 for a variety of safety violations. More important, it ordered SeaWorld to immediately cease all "water work," displays in which trainers swim and perform acrobatics with the orcas, because putting trainers in the water with orcas was inherently dangerous and not a controllable situation. The orcas, OSHA had decided, were actually in charge. SeaWorld bitterly assailed the ruling as "unfounded" and contested it in court. It lost its hearing before an administrative law judge and was fined $12,000 to boot.

A few right-wing voices had chimed in on Brancheau's death. Bryan Fischer, a preacher/pundit affiliated with the American Family Association, contended that Tilikum should have been put down permanently after he killed a trainer the first time and should be put down now, because the Bible commands you to kill any animal who harms a human: "Your animal kills somebody, your moral responsibility is to put that animal to death," he wrote. He also noted that the same Scripture dictates a death by stoning for any animal (and its owner) that kills a second time, but later clarified that he didn't think Tilikum should be stoned to death, just "euthanized, which can be done humanely."

That might have been a consideration with an animal of ordinary value, but Tilikum is worth several million dollars, and the thought of putting him down for killing a human never crossed the minds of his owners. Moreover, Tilikum's value to SeaWorld extends well beyond the raw market; he is, in fact, the cornerstone of the company's captive-breeding program.

During the OSHA hearings, it emerged that Tilikum was the top producer of semen that was then used to impregnate female orcas throughout the SeaWorld breeding program. Indeed, it was revealed that he was the father or grandfather of more than half of SeaWorld's captive-born killer whales. All of this was vividly documented for public consumption, first by David Kirby in his thorough accounting of the history of killer-whale captivity, *Death at SeaWorld*, and then, even more prominently and powerfully, by Gabriella Cowperthwaite's stunning documentary on the Tilikum saga, *Blackfish*. The movie, released the summer of 2013, was a critical and box-office hit, and suddenly, the issue of orca captivity was at the forefront of the national discourse.

SeaWorld had mostly played it low-key with Kirby's book and not responded to its publication, but when *Blackfish* hit theaters, the company struck back. It first tried attacking the film by claiming that it was "a dishonest movie," offering news media a list of eight points of rebuttal to the film's content. However, these points, in reality, were all distortions or factually false themselves, including the assertion that we don't really know how long orcas live in the wild.

When CNN aired the documentary on its network for the better part of two weeks in October, 2013, SeaWorld refused to even allow any of its spokespersons on air for any of the multiple on-air discussions that ensued. Instead, they offered a boilerplate condemnation rich with self-congratulation, calling the film "inaccurate and misleading" and suggesting it was mainly intent on exploiting a tragedy. It added: "Perhaps most important, the film fails to mention SeaWorld's commitment to the safety of its team members and guests and to the care and welfare of its animals, as demonstrated by the company's continual refinement and improvement to its killer whale facilities, equipment and procedures both before and after the death of Dawn Brancheau."

During a panel discussion on *Crossfire,* designated conservative Newt Gingrich attempted to take SeaWorld's side, but he had to acknowledge that he was "disappointed that SeaWorld isn't representing itself," adding, "I think, as a multi-billion-dollar institution, they owe the country some level of transparency and some level of accountability. And I am disappointed that they're not here tonight."

When you're being lectured by Newt Gingrich, a renowned backroom dealer, about a lack of transparency, and everyone knows he is right, you should realize you've lost the game, but SeaWorld never did.

Rather than back down, SeaWorld countered, in typically opaque corporate fashion, by creating a technologically flashy website titled "The Truth About SeaWorld," featuring the testimony of numerous current SeaWorld trainers and veterinarians either describing the wonders of captive life for their orcas or the supposed falsity of the testimony of the various ex-SeaWorld trainers whose testimony in the interviews featured in *Blackfish* was so damning regarding the company's treatment of orcas and its breeding program. The website claimed, for instance, that SeaWorld "does not separate mothers from their calves," which is true only as long as one defines "calf," as SeaWorld does, as a killer whale that is one year old or younger. Once they are over a year old, the company in fact transfers killer whales from facility to facility and separates mothers from their offspring in a manner that is completely unnatural, with relative ease and no apparent pangs of conscience whatsoever.

Most notably, the "Truth" website attacked *Blackfish* for supposedly blaming Dawn Brancheau for her own death, based largely on the ruminations of the various ex-trainers over the sequence of events leading up to Tilikum's attack. This was kind of an odd accusation, since the film not only documents two instances of SeaWorld spokesmen publically (and officially, before the OSHA hearing) saying Brancheau was at fault for her supposed mistakes but also shows the ex-trainers' outrage over these attempts to blame the victim.

However, shortly after the SeaWorld website went live, the CEO of Blackstone, the holding company that owns SeaWorld and took its stock public in 2013, went on national TV and blamed Brancheau again. Stephen Schwarzman, in a January 23, 2014 interview on CNBC, claimed that SeaWorld "had one safety lapse—interestingly, with a situation where the person involved violated all the safety rules that we had."

The next day, a PR firm hired by Blackstone corrected their boss in a subdued fashion, issuing a press release saying that Schwarzman "misspoke on the details of the death of SeaWorld killer whale trainer Dawn Brancheau. ... Mr. Schwarzman was unaware of the precise circumstances of the incident, which occurred nearly four years ago, and his comments did not accurately reflect the facts of the accident or SeaWorld's longstanding position on it. Dawn's death remains a source of great sadness for her family, friends and colleagues and Blackstone regrets the error."

However, they added that Schwarzman had no plans to return to CNBC to correct the record on air.

CHAPTER Six

The Guardians

I N 1967, ANIMAL PSYCHOLOGIST PAUL SPONG HAD BEEN HIRED TO HELP
care for Skana, a Southern Resident female orca who had been captured by Ted Griffin and crew near Yukon Harbor. A brash younger
version of the respected scientist he would become, he made the mistake of speaking up.

Skana, whose name was the Haida word for *killer whale*, had quite
the sketchy history. Captured as a calf, she had been sold to a boat show
that put her on display in a large plastic tank and fed her hamburger
and dead fish, which she refused to eat, so they then force-fed her. Her
health failing, the owner of the Vancouver Aquarium had stepped in

*Paul Spong checking his monitors' gear at the OrcaLab overlook on West Cracroft
Island.*

and taken Skana off their hands for $25,000, put her in a proper tank, and restored her health.

At the time, Spong, a native of New Zealand, was an assistant professor at the University of British Columbia (UBC), and the aquarium hired him to provide care and advise them on how to handle Skana, as well as another calf, a male named Hyak they acquired shortly thereafter. For two years, Spong did so, all the while recording the sounds they made and conducting his own behavioral experiments with them.

Skana was spunky. After spending a couple of weeks being fed fish for hitting the right response button on one of Spong's tests at a 90 percent rate, she suddenly and defiantly stopped and began intentionally hitting the wrong response button. "She was, in effect, telling me, 'Hey, I've got ideas and opinions, too, buddy. Don't think you can make me do whatever you want to do just because it means I'll get a dead fish,'" Spong recalls. Believing she was bored, he started introducing sounds into her environment and found that she responded best to music. However, she was easily bored and remained defiant.

Hyak, on the other hand, proved to be a real music aficionado. Spong found that he would perform water ballets to some classical music and breach and frolic to the Rolling Stones. He never seemed to want to listen to the same piece twice, sulking if the music was a repeat performance. Eventually, Spong found that he responded best to live music, particularly flutes, violins, and acoustic guitars: "He'd sit there so the point of his head was just millimeters away from the strings of the guitar. He'd be listening as if he were in ecstasy," Spong later recalled.

After a while, it became clear to Spong that captivity may be a great way for humans to get up close and learn about them, but it was bad for the whales themselves. So in 1969, Spong gave a speech at UBC, which was well attended and covered by the local papers, about what he had learned about orcas. "My respect for this animal has sometimes verged on awe," he told the audience. "*Orcinus orca* is an incredibly powerful and capable creature, exquisitely self-controlled and aware of the world around it, a being possessed of a zest for life and a healthy sense of humor."

Spong was well aware he would be accused of anthropomorphizing these animals, but was determined to relate what was his considered scientific view. So he continued with the next logical step, voicing his skepticism about the ethics of keeping orcas in captivity and ended that phase of his career in the process: "It has been my feeling, since observing the semicaptive whales at Pender Harbour, that *Orcinus orca* in the wild, in the company of family, is a decidedly different creature than the *Orcinus orca* we observe in the aquarium," he said, and he went on to suggest that aquariums look at providing their orcas with more of a semi-captive environment that might let them come and go at will.

The headlines the next day told the public that Spong thought Skana would prefer her freedom, and within a few weeks, he was no longer employed by the Vancouver Aquarium. (Skana eventually died of an infection in captivity in 1980; a wild Southern Resident male, L78, born in 1989, was given the same name.) He also left UBC—and academia—and took a $4,000 grant to study killer whales in the wild and moved with his family up to Johnstone Strait in late 1969. Spong set up a foundation he called the Killer Whale (Orcinus Orca) Foundation, or KWOOF, with the purpose to stop any further captures of live whales from Northwest waters. He also set up a shack on a bay overlooking Blackney Passage and began working with a handful of other researchers in the area, cataloguing the activities and, he hoped, the identities of the orcas who populated these waters.

This was the beginning of the first real human effort to act as guardians for killer whales, and it involved the scientific community that began the long and often difficult work of gathering data on these creatures. That core, once begun, has never ceased its work and is at the heart of what we now know about orcas.

• • •

Two years after erecting a shack covered with blue tarps, Spong bought the property it was on and built a full-sized laboratory and home there on the eastern point of Hanson Island, tucked into a little cove that offered decent protection from the weather and a terrific, open view of

Blackney Passage, where the fins of the blackfish could often be seen frolicking in the churning tidal currents. He called it OrcaLab.

Spong continued his experiments with music here and was perhaps a little dismayed to discover that wild orcas were, generally speaking, much less interested in his sound offerings, at least the recorded ones. So Spong decided to see if they liked live music better, and one summer, he brought in a live rock band to play from his boat. The response was much better, at least initially. The whales followed them at length, but no one was sure whether it was just the noise and novelty of it all that attracted them. At least everyone had a groovy time.

Spong also conducted experiments by playing his flute around the orcas from his kayak. It produced mixed results, but it made people wonder if Spong was a serious scientist. Indeed, OrcaLab became legendary as a fun place to hang out among the hippie folk and eco-activists who came to Johnstone Strait in the summers to see the whales.

All this was taking place in a time of great social ferment and shifting attitudes, coming on the heels of the Vietnam War and the rise of the environmental movement, one of the faces of which was a fresh appreciation for whales and dolphins and an increasingly critical view of the business of whaling and its effects on the world's wildlife populations. By the 1960s, scientists were warning loudly that the world's great whales were becoming endangered species, and this caught the public's attention, especially through conservation-oriented organizations such as the National Geographic and Audubon societies, as well as TV wildlife programs such as those produced by famed diver Jacques Cousteau. Pretty soon, "Save the Whales" stickers were popping up on the bumpers of cars driven by the environmentally minded.

On top of that, John Lilly and his vivid speculations about dolphin intelligence were gaining media attention, and dolphins and other cetaceans quickly became objects of fascination for the mysticism-minded. Even more stunning to public sensibilities were Roger Payne's recordings of humpback whales singing in an awe-inspiring chorus, which demonstrated viscerally to ordinary people that these were creatures with a special kind of intelligence. The recordings even made a best-selling LP that, of course, was a special favorite of the hippie set.

Paul Spong seized that moment to make it into something worthwhile. In 1975, he convinced Greenpeace director Robert Hunter, fresh off the organization's successful efforts to reduce nuclear-weapons testing in the open waters of the Pacific Ocean as well as on the tectonically unstable Aleutian island of Amchitka in Alaska, that saving the whales could and should be the environmental warriors' next great cause, and they decided that direct action would be the path.

Using his credentials as a whale scientist studying sperm whales, Spong obtained the coordinates of Russian whaling fleets operating in the Pacific. Shortly afterward, a Greenpeace team, including Spong, Hunter, and a fiery activist named Paul Watson, sailed out to intercept them, and on June 27, 1975, off the coast of California, they found a fleet of Russian whalers about to harpoon a pod of migrating sperm whales. A small fleet of high-speed orange inflatable boats operated by the Greenpeace crew zoomed up and interposed itself between the whalers and their prey. At one point, one of the Russian whalers fired his harpoon toward Robert Hunter and Paul Watson, but it sailed harmlessly over their heads. Nonetheless, footage of the whole affair was broadcast around the world, and it reached the public consciousness that humans had, for the first time in history, placed their own lives in jeopardy to preserve an oceangoing cetacean.

An international outcry soon followed, and then began to crest in the coming years as more whales were slaughtered and more protests arose; in the meantime, Paul Watson splintered off from Greenpeace two years after the Russian confrontation and formed his own organization, the Sea Shepherd Society, which remains active in confronting whaling operations. The reality that whales were becoming endangered species, and the pressure from the public outcry forced the International Whaling Commission (IWC) to begin rethinking its mission and enforcing real conservation of the world's vanishing whale populations. Eventually they won: In 1986, the IWC announced a global moratorium on all commercial whaling operations, and what whaling that exists today is conducted only by a handful of rogue nations that refuse to recognize the ban: Japan, Norway, and Iceland.

Even amid all the hippies and rainbows and unicorns, Paul Spong

remained serious about the work of collecting data and coming to understand wild killer whales better, and when the flakes had fluttered away
by the 1980s, and everyone's "Save the Whales" stickers had aged and
peeled off, he was still there doing the work. By then he was able to re-
cruit volunteers to collect data from a number of watch stations he had
situated around Johnstone Strait and some of the adjacent Broughton Archipelago islands. This involved, essentially, camping out at the shoreline
shacks, watching and recording the whales, both their surface behavior
and, with hydrophones, their underwater communications.

All along, of course, he was hardly doing this work alone. Spong was
only the best known of a coterie of scientists from various walks of life who
studied orca populations in Northwest waters and who opened our first
real window on the lives of these most intelligent cetaceans. Collectively,
all these people became the guardians of the Northwest's killer whales.

• • •

In the summer of 1970, when Paul Spong had moved there with his
family, Mike Bigg also showed up in Johnstone Strait to look for killer
whales, but he was a little better equipped. He was there for the government, so he showed up with a 40-foot boat and plenty of gear, notably photography equipment. Bigg was a London-born transplant, who
had moved to the Pacific coast as young boy and was raised with a fascination for the wildlife there. He obtained a biology degree from UBC
by specializing in the study of harbor seals, and in 1970 he had just been
appointed chief marine mammal biologist for the British Columbia unit
of the Department of Fisheries and Oceans (DFO).

Increasingly, the captures of the region's killer whales by marine-
park teams were becoming controversial with the public, and Paul
Spong's widely reported remarks of the year before had added fuel to
the fire. Worst of all, no one knew what the actual numbers of orcas in
Northwest waters were, information that would have to form the basis
of any management policy. The assumption was that there were many
thousands of the killer whales out there, so losing a few to captivity
wasn't likely to hurt. The DFO decided it needed to be sure.

So Bigg was ordered out to see if he could gain any kind of count of killer whales in Canadian waters. And he did. He printed out and hung thousands of fliers at boat docks, in coffee shops, and in fishing shops asking people to send or call in any information they had about killer whale sightings. The calls started flooding in. It was the start of his database.

Within a few months, Bigg had hit on the idea of using photographs of the whales, focusing on their dorsal fins and saddle patches, to identify them. He realized that if he could sort out individual whales from each other, an accurate census of the population would be possible. Bigg's methodology initially met with resistance and skepticism from scientists who were accustomed to more laboratory-intensive models of research that required captivity and who were doubtful that such nonintrusive research could yield much in the way of results. They were, of course, dead wrong.

Within a few years, Bigg was able to ascertain that there were in fact two distinct populations of killer whales in the Northwest, one that primarily resided in the inland waters of northern Vancouver Island in the summer and fall and another that spent the same months largely around the San Juan Islands as far north as Vancouver. More importantly, these whales numbered only in the hundreds, perhaps 300 at most. That one knocked everyone back, and it changed everything.

Eventually, after Bigg's numbers convinced everyone that the captures had to stop or there would be no more killer whales remaining, the governments of both Canada and the United States put a halt to the predations of the "orca cowboys" and placed their waters off-limits to would-be captors. Then, having accomplished that with finality in 1976, the DFO informed Bigg that his work was no longer needed and that his orca research was being defunded.

Bigg, of course, understood full well that the work was just beginning. So he continued to catalog the Northern Residents and their pods in the ensuing years, even without official funding. He managed to persuade the government in 1982 to set aside Robson Bight, the special spot in Johnstone Strait with the smooth-pebble beach where the

Northern Residents love to gather to rub themselves, as an Ecological Preserve, which meant that logging and development would be permanently precluded from the little cove, as would hordes of whale-watching boats.

In 1983, he was first diagnosed with cancer and, after enduring treatments, appeared to return to good health. However, in 1990, Bigg was told he had advanced-stage leukemia, just as he was wrapping up work on his final report on the Canadian killer whales. He read the finalized version in his hospital bed and died the next day, on October 18.

Bigg's legacy lives on in many ways: The Robson Bight preserve now bears his name, as do the "transient" orcas he first identified (not to mention the J pod whale named Mike, born in 1990). The fact that his photo-ID technique revolutionized not just cetacean research but the whole field of wildlife research—techniques that gave researchers non-obtrusive data-gathering tools for a broad range of management concerns—makes him something of an object of awe among his fellow scientists. And his bigger-than-life, gregarious, and generous personality remain deep in the memories of everyone who knew him.

"Mike's last name was right," says his old friend Ken Balcomb. "He was big—big in every way. Big heart, big mind. I still miss him."

Bigg's chief legacy, however, is the status now enjoyed by the killer whales he loved and studied: They are alive and surviving if not thriving, and they are widely loved by millions more people. They have become regional icons of the Pacific Northwest, talismans of its identity, and helping the Southern Residents survive is something of a fetish there. The local papers report assiduously on the births of new calves and the losses of old veterans, and orca fans keep careful track of the members of J, K, and L pods and their health. People know them by the names and numbers that Bigg first helped supply. That kind of heavy-duty public support for a conservation program made it easier for government officials on both sides of the border to take the steps needed to declare those pods protected under the Endangered Species Act in 2005.

The scientists who were Bigg's colleagues and companions are continuing to carry on his work, along with even newer generations of whale scientists and advocates.

• • •

Ken Balcomb is one such advocate. A native of California, Balcomb arrived in the San Juan Islands in the 1970s with a contract from the National Oceanographic and Atmospheric Administration to collect information on killer whales. Soon he and Bigg were working together to compile a complete census of the Southern Residents, which they completed in 1976.

"Our mandate was just, count 'em up," recalls Balcomb. "But within three or four encounters, I saw that it was worthwhile continuing our study, looking at whales growing up, to see how fast they grew up. Basically, it's like having all the fish in the fish bowl that you can look at.

"I wanted to ask more lasting questions that were of interest to other biologists—how long do they live, how many babies do they have, what is their behavior. And as it turns out, they are a society, with a culture and intelligence."

Ken Balcomb

Balcomb became enmeshed in the fight to protect the Southern Residents from captures. It was that '76 census, showing that only about 70 whales remained in the entire community, combined with Bigg's similar work with Northern Residents, that led to the shutdown of Washington and Canadian waters to orca captures.

Balcomb stayed on San Juan Island. In 1985, he bought a piece of waterfront property on its western side, a place where whales are frequently seen, back when such homes were affordable, and converted it into his Center for Whale Research (CWR). In addition to his own work conducting a variety of research projects, the CWR, from a two-story house overlooking Haro Strait, with a work lab up the hill, hosted summer-long gatherings of Earthwatch (an environmental organization) volunteers, who would go out on Balcomb's boats (usually a big trimaran that had been donated to the center years before) to observe whale behaviors, take photographs, and record sounds. In between, Balcomb became involved in a broad range of whale-research and rescue efforts, including dealing with beached beaked and minke whales in the Bahamas and humpbacks in the Atlantic. He's also played a key role in building orca census records in other parts of the world, notably in southeastern Alaska.

Nowadays, a little grayer in the beard, he is more retiring, content to maintain the annual census of Southern Residents and monitor their well-being with the help of his core team, but there are always projects popping up that need his attention. Over the winter of 2012-13, he spent much of his time pursuing the K pod down to Monterey after National Oceanic and Atmospheric Administration (NOAA) scientists successfully satellite-tagged one of its members.

• • •

Like his old friend Mike Bigg, John Ford had grown up in British Columbia and was entranced with its wildlife. As a grad student, working under the tutelage of Bigg, he had studied the vocalizations of Northern Residents for his master's thesis. In addition to his early work establishing the existence of dialects in killer whales' communications,

Ford did much of the leg work amassing the data on the Vancouver Island orcas who formed the basis of Bigg's completed survey. In the process, they spent many weeks and months huddling under tarped tents in the gully-washing Johnstone Strait rains, but when they were done, the science was impeccable.

Ford had early exposure to killer whales beyond even his first frightened encounter in his family's sport-fishing boat. Because his mother volunteered at the Vancouver Aquarium, he was afforded a front-row seat when Moby Doll was captured, and his imagination was fired. He got a job sweeping up after shows at the Aquarium, watching their captured whales, Hyak and Skana, who were initially kept in a small dolphin tank until a larger facility was made for them.

"The bigger pool was built in 1972," recalls Ford. "That was around the time I went to work at the Aquarium. I was initially a pop-corn sweeper. My beat was the stands around the pool, and so I spent a lot of time leaning on my broom and watching the whales. I was, of course, totally intrigued with them, and the next year I got a position in the marine mammal department—first, feeding seals and then belugas and worked my way up to working with the killer whales: Hyak first, and then Skana eventually. She was a real handful, so the experienced trainers knew she was a little ornery."

Ford kept the job through college. "It was kind of a summer job as I got my undergraduate degree at UBC. I left there in 1976 to work on my thesis, which I did on narwhals. Then I worked on gray whales in 1977. I switched theses back to my first love, I guess, after a particular dull and challenging encounter with gray whales, which was interrupted by some transient orcas—that convinced me I was going down the wrong path.

"So I went back to my initial desire to study killer whales, which I had approached Mike Bigg to do a couple of years earlier, when some of the first results from Mike, Graham, and Ian McCaskey's work coming out showing that these pods were really stable and could be predictably found. And I thought, perfect! This is an opportunity that I don't know that exists for any other species where you can repeatedly go out and actually find the same group over and over again over a period of weeks and months, and observe different behaviors and record their sounds, which could match up sounds with behaviors, and maybe get some insight into how those sounds function.

"Now, I'd been listening to recordings that Dean Fisher, my advisor at UBC, had of killer whales and he had some recordings of Moby Doll, and Skana and Hyak. I approached Mike on that, and I think he was a little skeptical—I even proposed to him that, you know, maybe they have dialects between different groups. And I think at that point he probably wrote me off as a wild-eyed guy. And it was true that I was naïve in even proposing that, because I didn't have sufficient background in animal communications at that stage to realize that that was a little outlandish to even suggest, because land mammals don't have dialects at that kind of level. Humans do, but no other species does."

Eventually, Ford would not only go on to prove that killer whales have dialects in their communications, but that these dialects adhere closely to the animals' social organizations. Bigg himself would ruefully recount the tale of his skepticism to his fellow scientists: "Mike always used to like telling that story about how important it is to remain open-minded because this young kid came over and proposed that he wanted to record whales because they maybe have dialects, and sure enough they did," Ford says with a smile.

After obtaining his PhD in biology from UBC, Ford taught there for a number of years while conducting orca surveys with Mike Bigg and others in Johnstone Strait. Ford eventually went to work for the Vancouver Aquarium, where he was in charge of caring for its captive orcas, initially, the same Hyak who had been under Paul Spong's care, and then later, two Icelandic captives named Finna and Bjossa. (Skana had died in 1980, after 10 years in captivity; Hyak died in 1991, having survived 23 years.)

Ford's most important work, however, involved wild whales. He initially specialized in analyzing the communications used by killer whales. During those years at the Aquarium, Ford conducted pioneering studies of Northern Residents analyzing orca dialects and echolocation, as well as research about their diet and social organization (the latter was especially an outgrowth of his important work demonstrating that calls from specific pods were signature calls that could identify them and that these calls could be taught to and mimicked by other orcas).

Operating in the Northwest, however, the Vancouver Aquarium was faced with increasing hostility from the local public about its captive whales, especially after Finna died in 1997, and a dilemma arose in deciding whether to obtain another orca to keep Bjossa company. They decided to punt, getting out of the captive-orca business altogether. They sold Bjossa, their last remaining killer whale, to Sea World in San Diego in April 2001. She died there six months later.

By then, John Ford had already found other work; the Department of Fisheries and Oceans hired him in 2001 as head of cetacean research for its Pacific Biological Station in Nanaimo, an hour's drive north of Victoria on Vancouver Island. From there, he continues to conduct study after study. Unlike most of the scientists who study orcas in the wild, Ford's background inclines him to mixed feelings about keeping whales in captivity.

"My feeling about captivity has evolved over the years," he says. "I was never wholly accepting of it—I could accept it in certain institutions that had a strong research or at least an education focus that treated them with a degree of respect. I've never been a fan of the highly theatrical shows involving water work—I just don't think it's portraying the animal appropriately, and I'm not sure that the audience is getting a message that really contributes to the betterment of killer whales, toward their conservation or anything like that.

"So most of the institutions that now house killer whales are more of the corporate kind involving the more theatrical shows, and I don't think that really is an appropriate kind of environment for these animals. I think compared to other species, they don't do as well in captivity. It's really difficult to provide for their physical needs and social needs.

"That said, if they're not available in the future for research opportunities to learn about things that we just can't get a grip on in the wild, that's going to be really unfortunate. Especially when it comes to physiology and some of the acoustical stuff. But the kinds of facilities that house killer whales today and the way that their show schedule dominates their time, it doesn't really provide opportunities for that kind of research."

. . .

Graeme Ellis knew Bigg, too, but unlike his good friend Ford, he is open in his contempt for orca captivity. He had grown up in Campbell River, just a few miles north of Nanaimo and a little south of Johnstone Strait, but well within the range of the Northern Residents, and he would see them while growing up. He too got a job early on with the Vancouver Aquarium, caring for a group of wild orcas who had been captured and were penned up at Pender Harbour, adjoining Georgia Strait from the mainland.

Ellis came to know several of the whales during the time they were in his care. Eventually, he tried getting into the water with them in his wet suit. One of them, a big male, came charging at him, teeth flashing, and Ellis leapt out of the water: "He scared the hell out of me. One moment I was in the water, the next moment I was standing on the logs."

But then he tried it again. This time, the big male charged again but stopped short of Ellis and then nuzzled him like a big dog. It had all been a big bluff. Ellis scratched the male under his chin and along his back with his fingers, and the orca nuzzled him some more. They became pals, at least until the day the big male also proved to be clever and escaped back to the wild.

Ellis was eventually laid off by the Vancouver Aquarium, but the experience helped him land a job caring for orcas at Sealand of the Pacific, a large sea pen on the Haro Strait side of Victoria that was designed like a cattle yard for killer whales. The orcas were kept in increasingly appalling conditions as the years went on. Among Sealand's more notorious facilities was the set of small enclosures in which the orcas were all stored overnight, allowing them no freedom of movement at all.

As a trainer at the facility, Ellis swam with its whales often, particularly a big male named Haida. Ellis says Haida once gave him a kind of warning dunk after Ellis had failed to take the hint that playtime was over, reminding him, once again, who was really in control. It made clear to Ellis that orca facilities are only possible because of the patience and forbearance of the orcas themselves. Ellis was also among the Sealand crew who captured the albino whale Chimo in 1970, along with the rest of its T2 family, some of which later returned successfully to the wild, the same crew who misfed these transients a fish diet for the duration of their captivity. By then, however, Ellis had undergone a change of heart.

"I got really disillusioned," he said. "The whole thing was money-oriented. You sell the whales, you sell the show, and if the whales don't perform, people don't want to see them. As a trainer, I had people demand their money back because a whale wouldn't jump if it was off its show. I got quite cynical about the reasons the public was there." What really appalled him, he said, were Sealand's goofy entertainment shows featuring orcas wearing hats and sunglasses and being ridden like horses, ending with the requisite soaking of the audience. "Those shows just made me want to puke," Ellis said. "It was degrading to the animal. . . . I thought it was just sick."

Ellis took time off and then went to work for Mike Bigg. It was Ellis who became Bigg's ace photographer, forming much of the foundation of his photo-identification database (which included some 7,000 photos). Over the years, Ellis worked with the DFO as a technician gathering data on marine mammals of all kinds, as well as tracking salmon movements and assessing shellfish populations. Eventually he took a position as a researcher at the Pacific Biological Station in Nanaimo, where he remains to this day, working with John Ford specializing in work with whales and large marine mammals. In recent years, he has conducted a wealth of research on transient orcas, Bigg's killer whales, and analysis of the world's populations of killer whales.

• • •

Alexandra Morton, like most of the Northwest whale scientists, knew Mike Bigg well, too, but her pedigree working with killer whales began farther down the coast, in California; Morton at one time was a protégé of John Lilly, the renowned dolphin scientist whose reputation later fell into disrepute after he tread into dubious experimental methods (such as giving LSD to dolphins) and held strange speculative views. It was Lilly who, in 1977, employed a young Alexandra as a researcher at the Human/ Dolphin Society, where she catalogued some 2,000 recordings of dolphin vocalizations.

The next year, she went to work at Marineland of the Pacific in Los Angeles, first working as a dolphin trainer and conducting research on their communications. She then befriended the park's two orcas, a pair of Northern Residents named Orky (a male) and Corky (female). The two whales kept mating, and Corky kept getting pregnant, raising hopes at Marineland that they might someday be able to produce a viable captive-born killer whale. However, it never happened; Corky either miscarried, or her young calves would die shortly after birth because, having been captured in 1969 at only about three years of age, she lacked the knowledge or skills to nurse the youngster.

The first of these lost calves was particularly heart-wrenching for everyone involved. Morton described it in her anthem to orcas, *Listening to Whales:*

> I had learned a few orca calls while recording Corky's false labor, but the sound that Corky repeated the night she lost her baby was new to me. This wasn't a sweet rising and falling riff. This was strident, guttural, and urgent, like a dog yelping at the end of a chain. After each breath, Corky returned to the bottom of the tank. As her delicately curved face grazed the cement bottom, she resumed her lament. Every fiber of her being begged for the nuzzling of her newborn baby.

Alexandra went to work with the two orcas, believing she could make a difference in helping improve their lives. However, she realized it was futile, especially after observing Corky sink into depression. "It ignited the awareness in me that captivity had a certain wrongness about it, and that my race had violated this mother whale in the most basic sense," Morton later said. After a year at Marineland, she gave up in anger and frustration. Moved by concern for the orcas she was leaving behind, she contacted Mike Bigg to inquire about the families of Orky and Corky. "Mike just blew me away," she recalled. "He was extremely cooperative, offered all the help he could."

It convinced Morton to come north to Vancouver Island and OrcaLab, where she went to work as a volunteer researcher. It was while there that she encountered a man named Robin Morton, a Canadian filmmaker who led a crew shooting underwater footage of killer whales. That summer they were trying to obtain footage of orcas rubbing themselves on the smooth pebbles of Robson Bight. Alexandra and Robin fell in love and were married in 1980, then moved to the Broughton Archipelago and set up shop in the outback village of Echo Bay, which is mostly a collection of homes with docks. In 1986, Robin Morton died in a diving accident; his rebreather stopped functioning while he was deep underwater at Robson Bight. Alexandra Morton, however, picked up the pieces and continued their work as advocates for the killer whales. She remains in Echo Bay to this day, having remarried and raised two sons in the quiet wilderness. In recent years, she has become a persistent critic of the salmon-farming industry in British Columbia, primarily because their operations pollute British Columbia waterways where native salmon runs are affected in a variety of negative ways. That in turn, she says, is harming resident orca populations.

• • •

You can learn a lot from whales, just trying to see them in the wild—not about the whales, but about yourself—because in trying to enter their world, you have to leave some of yours behind. My wife Lisa and I, both having been raised inland, first attempted to see killer whales on our honeymoon in 1989, when we traveled north to Alert Bay in British Columbia and signed on with local guides to spend a day sailing with the orcas. It was early July, and the whales had been plentiful. As it happened, the day we went out with them was the first day on which these guides encountered no whales. We were mildly disappointed, but we knew that killer whales could also be seen in the San Juan Islands, and having moved that year to Seattle, we figured it would be a simple matter to remedy. The next summer, we began making trips to the islands, always with the idea in mind of seeing orcas, but somehow, we kept missing them.

Seeing them was always more or less an agenda item, something we wanted to do while out visiting, but on our own schedule, and the orcas never managed to mesh with it. We would drop out to Lime Kiln Park or San Juan County Park and expect to see them from the island's western side. When we'd get there, inevitably, our fellow whale watchers would inform us we had just missed them. Or they just wouldn't show up that day. After a while, Lisa began joking that we were jinxed, that the whales were actively avoiding us and would flee if they knew we were coming. I wasn't so sure it was a joke.

Finally, after about three years of almost frantically driving about the island in the vain hope of catching a glimpse of the killer whales, we gave up, sort of. We began to realize that the places where we were looking for them were unusually beautiful in their own right and that there was more to see there than just whales. So we settled down and began enjoying the places themselves, bringing our picnic basket and blanket and books out to the west side, settling down, and just enjoying the day. At first, this meant simply that we were enjoying the waves and the wildlife and the gurgling quiet of the landscape.

Then we began seeing orcas. Lots of them. Our first spectacular view of them came from an area on the west side of the island known as the "Land Bank," an open space south of Lime Kiln where the orcas often come past, available for simply sitting-and-watching by the public. They breached that day, and there were calves romping next to shore, and the whole experience was breathtakingly memorable.

After that, we began taking more time and care, ensuring that we simply gave ourselves time to just be where the whales were, letting them come to us on their own terms in their own good time. We began camping out on the island and then kayaking. Most of all, we learned to just *be* where we were, in the whales' space and on their schedule, which was bound up in the flow of the tides and the salmon. If we did that, they would always come and bless us. Almost always, anyway. No

Wild orcas inevitably reward the patience of would-be admirers.

matter how well you master this, the orcas will always keep you guessing, but even when they did not, we had long since learned to soak in the beauty and stillness of where we were.

Monika Wieland went through the same sort of initiation. Nowadays, the thirty-something blonde is something of a San Juan Island wildlife icon, having published a masterful collection of photographs of killer whales that is popular in the bookstores, as well as a useful guidebook on bird life on the islands. However, back when she arrived here in the 1990s, she was much like Lisa and me: running hither and yon, clutching binoculars and cameras, hoping for some fleeting glimpse of the elusive orcas.

"I was definitely a whale chaser, going along the west side and trying to catch them," she says with a wise smile now. "More often than not, it doesn't work out."

She went to work on the tour boat *Western Prince* as an onboard naturalist, explaining orcas and the other wildlife to passengers, and in that way got her fill of whale sightings (not to mention a refined eye for wildlife photography). Then she left to take up other work on the island and found herself jonesing for orcas again.

"As I stopped working as a naturalist, it was really tough to transition, to go back to being an exclusively shore-based whale watcher, and not going to the whales, but sort of readopting that mindset," she says. "And when I finally got there and let it go, sort of, like: 'No, I'm just going to go to the west side and have a free afternoon and whatever happens, happens.' And just sit at Lime Kiln. And some of the best experiences happen when you do that. It's amazing how true that is."

In spite of all the time she's spent being around and observing wild killer whales, she says they remain mysterious to her. "I've been lucky to be here for as many years as I have, but I also feel like I'm just beginning to come to an understanding of the whales, in a lot of ways. They certainly are continually surprising me. I think I am starting to understand what they might do next, and they don't. They do something completely different."

Wieland says she has learned a lot more about herself, being in the orcas' world, than she ever thought she would. "I guess one thing that the whales have taught me is the benefits of just sitting in one spot and waiting for something. So much of our lives has gotten so fast-paced, and scheduling every moment and going from this to the next thing, and all the electronics and everything. But when I come to the west side, I sit outside, and I stop everything else, and go on whale time.

"I see amazing encounters out of that. There's a lot of other amazing things, other wildlife—the seals, and sea lions and the bird life. And also the people that I've met while waiting for whales has been amazing, too. I mean, I've formed some lifelong friendships doing this, sitting on the rocks, waiting for whales.

"In modern-day society, people don't wait for much of anything. But there is nothing instant-gratification about the whales, unless you're really, really lucky. Being on whale time means slowing down and looking around, looking at the people that are around you—what are they experiencing, and why are they here? And what are these birds? And the tidal life. There's so much to see."

Bob Otis has been joining Wieland on the west side of San Juan Island for many of those watches. A quiet-spoken, semi-retired college professor from Wisconsin, Otis mans the whale-watching operations inside the Lime Kiln State Park lighthouse, where a number of students and volunteers each summer monitor the comings and goings of Southern Residents and other whales who happen past. The monitoring work includes listening in on the hydrophones that are plugged in off the rocky shore at the lighthouse. Otis began this work in the 1990s as part of his work as an animal psychologist. But after a while, he says, it just became an old-fashioned passion: "Certainly when I started, I came with all the baggage that a scientist brings in terms of quantification and objectivity. That has changed somewhat," he says. "Today, when I teach a course on the killer whale, back in Wisconsin, I bring the students here," he says. "I want them to be able and willing to look at the killer whale through the eyes of a poet, a musician, an artist, as well as a scientist. I encourage my students to dabble in the arts, because it makes them much better scientists."

More than anything, Otis says, the orcas themselves have inspired him to think beyond just the raw science. "I think Paul Spong has said over the years that he started out being an objective scientist, and always stressed that kind of information, but at the same time he was willing to talk about personality differences in the whales and so on. And I think the same thing has happened with Ken Balcomb," Otis says.

"So is it the whale that does it to us? We hear from hundreds of people here every day who come in, and they perceive the killer whale as the icon of a perfect world. Anyway, it's very satisfying and rewarding. And I think if you hear that a few thousand times as a scientist, you start to feel that somehow our objective measures are not tapping into that, and they should."

Otis says that whales have rearranged his views about scientific certainty. "Certainly, the more we learn about these whales, the less we understand," he says. Moreover, his experiences have taught him to learn from unexpected sources. "I tell a story when I'm giving my little slide show about the whales breaching, about a little boy who was in the group and who raised his hand," Otis says. "He said he knew why whales breached. At the time, I thought he was setting me up. So I said, 'OK, why do they breach?'

"He turned to the group in the room and spoke clearly and, in all sincerity, he said: 'They breach so they can dry off.' And you know, he just might be right. It may well be something that simple. I have learned not to argue with the wisdom of children."

Indeed, Ken Balcomb has experienced much the same sort of epiphany in his dealings with the whales themselves: "They always seem to find ways to surprise me," he says. Mostly, he says he is endlessly impressed by the orcas' own flexibility and willingness to learn, as well as the depth of their empathy.

"The most amazing stuff was when A73 (Springer) was down in Puget Sound, and Luna up in Nootka Sound. I had gone to see Luna with Graham and John in December, after the marine mammal conference, and he was definitely seeking human contact. Not more than ten days later, Mark Sears called me and said we've got this little lone orca down in Puget Sound. I went down there and went out with Mark; the day before when he had seen her, she was playing with a stick, so we boated out there and found that stick. And there was Springer playing with it. After a while Mark would take the stick, and she would respond. I gave her a signal, and she responded in kind by rolling over. When it came time to capture her, she just let us do it. She didn't fight at all. She was ready for anything that we did."

He still is trying to grasp the role that killer whales play in their environment: "Maybe they do like the First Nations people thought they did at first—they go out and find the fish and herd them in. And they know enough not to eat all of them. In fact, they only eat about 10 percent of them. They know that it just doesn't work, if you eat them all. And whether or not they experimented with that eons ago, I don't know, but certainly in their management of things, they are very, very conservative. They are also conservative in their behavior in that they don't risk their lives.

"I wouldn't be surprised if there is some kind of stewardship involved. They've been very, very successful for a very long time."

At times, he says, it has seemed as though the orcas knew that he and his researchers were trying to gather information on them to help them and were eager to help them learn: "In the case of mothers and calves, we had mothers and calves rolling up and showing us their undersides. I think they were aware that we were curious about what was going on, and they were aware that knowing what sex a new baby was would be part of what would be interesting to our little monkey minds."

What remains with him in all his dealings, he says, is their similarities to humans: "They have a sense of humor," he says. "They play games and are clever. They outwit us and know it. It's very easy to anthropomorphize with these guys. Because as profoundly different from us as they are, they are also like us in a lot of ways. And that self-recognition is both shocking and inspiring."

Paul Spong, too, remains impressed by killer whales after all these years of being around them. His abiding impression is of the sophistication and power of their cultures.

"Their culture certainly seems to be real," Spong told me one summer afternoon at OrcaLab, looking across Johnstone Strait toward Robson Bight, the Northern Residents' favorite rubbing beach.

"You know, if you look at the different populations of orcas around the world and just take note of the fact that they have very real traditions, using the word culture does seem appropriate. The culture of the residents here seems to include this rubbing phenomenon. It's an amazing phenomenon, especially when you see the enthusiasm that you see little kids picking it up, it's really quite, quite remarkable." He laughed his gentle laugh.

"It's in a way unique to a particular group in a particular place, although the more that we learn about them, the more we find there are other rubbing spots that they enjoy, we just happen to know about this particular one. It does seem to reflect a longstanding tradition, which is unique to this group, so I think it's appropriate to use the term culture with respect to them.

"You look at the other groups—down in Patagonia, the orcas that strand themselves intentionally on beaches in order to catch baby elephant seals. Certainly, they are in a sense, effective hunting techniques, but it's so extreme—I mean, do they really have to go to this in order to get a bite to eat? I rather doubt it. There's no particular—in order for the dietary side to be satisfied with this hunting technique—there's no particular reason why they have to do this intentional stranding. It seems to be, for them about the challenge. And within that culture, that's what they do. So culture is an appropriate term."

Spong has no abiding doubts about the intelligence and sentience of killer whales. "I think they certainly seem to be very well organized, at a bunch of levels: socially, certainly," he says. "It's obvious that the brain gives them the capacity of organizing and dealing with their life in a coherent and conscious way. And they have a very successful life. They've been around in this form and doing this thing that they do, organizing societies with obvious cultural periods, for very long periods of time. This is just a phase in a much longer history. The modern form of orca has been around for hundreds of thousands of years, if not millions. The distinction between transients is something that happened back seven hundred thousand years ago. So you're dealing with really long periods of time.

"The bottom line there is if they are capable of being self-organized for such long periods of time, we ought to let them go on doing it."

John Ford respects killer whales for being creatures whose importance to the world is not defined by anything human. "They've been around a long time, they're highly successful, certainly in terms of the geographical scope over which they live, the range of habitats from the polar seas to the tropics—it's amazing," he says. However, he studies them because he fears for them: "They are vulnerable to extinctions, and the populations are not very large. But that's typical of apex predators, whether they're killer whales or lions or tigers. They are never all that abundant, and so they are vulnerable.

"In the case of these animals, certainly their societies have been evolving for millennia, and they are far more sophisticated than I think we know, or give them credit for, even though we do admire them, and I think most of us are highly impressed with their capabilities and their adaptability. The big concern is that we are having impacts on their environment at a global level, and so how resilient are they for some of the threats that are down the road for them?

"This is why I think it's so important that they are not so much the ambassador species, but they are iconic—people are naturally curious about them and admire, some people are obsessed by them. Whatever it takes, keeping them in the public's mind and eye is going to be important for their future survival, because we humans as a species are going to have increasing impacts on that—either through their competition for food, or the more insidious global issues, like ocean acidification. We're at the point now where humans can affect the fate of all other species. And this is a species that I think deserves much more attention than they're currently getting in terms of understanding their life history, their culture, and so on, and understanding what they need to survive going forward."

For Ford, there is a personal component: "On many fronts, these animals have played such a big role in my life. And I'd like to continue doing it as long as I can still walk down the dock to the boat," he says.

But he is most struck, and amazed, by the sea change in the human perspective on killer whales in recent years, coming at a critical moment in their survival and that of the entire oceangoing marine ecosystem: "I guess we've seen, in my lifetime, an incredible transformation of the animal in the mind of man—from vermin to icon, in what I consider to be my fairly short lifespan to date. It's quite dramatic, and it just never ceases to amaze me how we treat them so differently now.

"And yet the other thing that is astonishing to me is how we have systematically abused all these species of marine mammals, mostly through direct culling, but also through killing for profit, culling them just because they eat too much of what we consider to be our own fish. It was just one after another—first it was the sea otter, then it was the elephant seal, and then the sea lions—machine gunning them and hunting them out. And then the whales—first the large whales, and then later the others. Very few marine mammals got away without being driven down, for various reasons, by humans. Even killer whales were affected, through the live captures, and prior to that probably through pretty widespread directed shootings; they were just despised by mariners. It's just astonishing to me now the way our species has treated these animals, mostly since the arrival of Western civilization to our coast here, and the improvement in those attitudes now—we wouldn't think of machine-gunning sea lions any more, or even catching a killer whale for display anymore in this part of the world—it's sort of unthinkable. Which is a good thing.

"But now, the pressure is not off these animals. We are affecting them in more insidious ways, through things like pollution, through things like competition for food, and so on. So we need to keep the focus on them for their future well-being and ours. If we know we care so much about the whales, and we can't save them, then what's the chances for species that aren't in the public eye so much? That's why they are so symbolic of the health of the oceans in this area. If we can ensure that the whales do well, then everything else will probably fall into place, because the apex predators sit atop the food web."

Brad Hanson, who has been studying orcas since he was a child and grew up to become the chief whale scientist for the National Marine Fisheries Service's Seattle region, knows all about their magnetism. He was around when Namu came to town, and he was among the entranced children who lined up at the waterfront to see him. "I knew then that I wanted to work around these guys," he says.

The thing that killer whales have going for them, Hanson says, is that unlike other animals—say, salmon or frogs—orcas are breathtakingly charismatic. "I think people identify with them on different levels. One aspect is the black-and-white animal thing—there's a sort of sea panda aspect to them. Visually, they're striking.

"They're also a top predator. We're a top predator. I think there's in some people's minds a mutual respect. They were feared for a long time. Just like people have a fascination with sharks, people have a fascination with killer whales. They're these big, powerful animals.

"But even beyond that, what we've learned in the last forty years, of course, is that they're very social, so they're a lot like us. And I think it's the similarities to us—that they're social, they have a culture, they communicate with each other—it's all those things. They have a lifespan that's almost the same as us. If you looked at that aspect alone, there's very few species—maybe elephants—that have very similar traits to us. And I think that humans are fascinated by that."

"And just the way they lead their lives, and the nature of their long-term ecological success as a species, should provide lessons for human beings," Hanson says. "One of the things that does fascinate us is that these guys all seem to get along pretty well. Which is unusual for most mammal species, including ourselves, given how much time we spend at war with each other. They're very strange in that regard, in particular because they don't kick males out. Usually, in mammalian species, males are gone once adolescence kicks in. Resident types don't do that. The fact that they do get along—I'm not sure we can learn how they get along, but it's the fact they do get along is something that's of value to observe in the natural environment."

For Fred Felleman, one of the killer whales' fiercest advocates in the political realm—his Orca Conservancy played a vital role in reuniting Springer with her Northern Resident family—the lessons provided by the example of killer whales is vital for the very same humans who are fighting to sustain and help them. Felleman came to Seattle in the early 1980s to get a biology degree and stayed on to organize the environmental and tribal groups who have played the central role in getting the Southern Residents listed under the Endangered Species Act.

"Because I spent all this time getting my degree and in most of the time since, I've been dealing with ships and whales," Felleman says. "And my clients are these interesting, mostly tribal and environmental organizations. And they're not what you would call the most lucrative clients, but what has been remarkably obvious to me—and from the outside, you would think these tribal governments have so much more in common than they do as compared to them and us—is that in unity, there is strength, and you would figure they would bind together to try to take on the greater challenges of being a red man in white country.

"And similarly, from my other clients, the environmentalists, we've learned that we will never be able to outspend the big boys, and we're never going to be able to have the same kind of lobbyist interests in the halls of Congress or the Legislature, and that again, we would do best to coalesce around the cause and try to make up for our lack of financial resources with numbers.

"So having spent the past twenty years or so doing this, what I realized is that one of the hardest things for human beings to do is to cooperate, and that's the most important thing that the killer whale can teach us, as a cooperative hunter: There is unity in strength. They're not the biggest son of a bitch in the valley, but they are the top predator. They are the baddest, as it were. And as a model of how to make a living in a world where there are giants around you, the killer whale has one of the most important lessons to teach us."

Val Veirs, a retired physicist who has been listening and gathering orca sounds for several years, believes that breaking the communication barrier, as unlikely as it might be, could open opportunities for human education not just in relation to orcas but in relation to the natural world they inhabit. "They do have a certain kind of long-term memory because they know how to organize their lives around the comings and goings of things that are not easy to understand," Veirs says. Presumably, that's why the mothers live so long, because they pass on geographic and other wisdom. "So maybe they have some long-term wisdom that if you could communicate with them, they could say, 'We have an oral tradition in our world, and our oral tradition says that at one time you couldn't catch any fish here, because it was ice all over the place, and then the ice went away, and then the fish started to come up here where the fresh water is.' Maybe they've got all that encoded."

There might be lessons simply in long-term survival there, too, Veirs says. "Why are they so specialized?" he wonders. "Given that they don't have any real competitors, maybe ecological niches are a way that species can carve out a way to avoid competition. Maybe that gives you a certain kind of stability that allows, say, the residents to coexist with Ts without competing with them."

Howard Garrett, who has been watching whales ever since his brother, Ken Balcomb, brought him out to the Northwest in the early '80s, has been applying his background as sociologist to his observations. Garrett says that "symbolic interactionism," the notion that people act towards things based on the meaning those things have for them, and those meanings are derived from social interaction and the symbolism that attaches to that, goes a long way toward explaining killer-whale behavior.

Garrett contends: "Orcas and humans have independently evolved (or converged) into symbol-using animals. Behavior based largely on symbolic interaction fosters the creation of complex cultures that allow action based on meaning, interpretation and choice, which can, and usually does, override genetic or environmental factors in determining behavior and speciation." He writes that "self-recognition, role-taking, interpretation of the generalized other and use of symbols are all essential precursors for the development of culture, whether in orcas or humans."

There is a lesson for humans, too, Garrett says, in *how* orcas organize their societies: "I think it's live together and enjoy life. Learn to transcend hostility. They're as capable of it as any creature on the planet. They have hierarchies of dominance, beginning with the matriarchs, but there's no sign of discipline, there's no jousting for position. You only see occasional scuff marks or rakes, mostly on young ones. But they don't butt heads, they don't beat each other up.

"One of the few universal behaviors that they have is that they do not harm humans. They are unique among apex predators that way. And that alone stands as reason to heed whatever lessons they might have for us."

Dolphin neurophysiologist and ethicist Lori Marino, too, has speculated about what killer whales might be able to teach us, should we ever breach the communication gap: "One, I'm not sure we ever really will," she warns. "But you would hope that what you'd want to learn is something about who they are, and maybe something about whether they share an internal world—their phenomenology, or their subjective experience. I think just the idea of being able to have communication with them that we can confirm as communication—right now, we do communicate, but we don't speak the same language so we never really know—but the idea of just knowing that there could be an exchange where both parties willingly know what is going on, I think that would be quite amazing.

"I don't know if we will get there. I just think that their communication system is probably more complex than we can imagine. I think we could get closer, but people have been trying for decades, and we still haven't really figured out much."

Nonetheless, of all the animals she has studied, *Orcinus orca* is the one that most inspired Marino's awe. "They were the brainiacs of the planet well before we were," she says. "The thing is, one of the things that you can learn from cetaceans is that there are ways to survive when there are many different types of species of the same kind. I mean, we're so used to putting ourselves apart from the great apes, we're the only hominids around, and that gives us a false sense of who we are. The seventy-seven species of dolphins and whales and porpoises have been sharing the ocean for tens of millions of years, and managed to do that.

"Look at the orcas: They manage to partition resources, even when they live in the same region. So they've figured that out. They've done it better than we have. But it tells you that if you have a big brain and you're really complex, then you can do it.

"We went off that track, separating ourselves from nature, thousands of years ago. And I think part of the problem is that when you don't allow yourself to be connected, then you can't expect to really understand other species. Part of not understanding orca communication is the presumption that it's got to be much less complex or less deep than our communication. And once you start from that point of view, you're not going to get very far. But if you start from a different place, you just might."

CHAPTER *Seven*
Salmon, Boats, and Oil

THE SUMMER OF 2013 WAS A PECULIAR YEAR—A WORRISOME YEAR— for whale watchers in the San Juan Islands, and it all came down to salmon, most likely.

Simply put: No salmon, no whales. The resident killer whales just were not around much at all in the Salish Sea that summer. The J pod and some K pod whales were around in the first week of June and then pretty much vanished. Some J pod whales briefly reappeared in mid-July and then promptly returned to sea. They did not return with any regularity until mid-to-late August.

This was most peculiar. In all the years they have been carefully observed here, the resident orcas have always been plentiful in late June and early July, with numbers slipping into August and September, but in 2013, those weeks were barren of orca sightings.

Large ships and their noise loom almost constantly in the lives of Southern Residents.

There were those three L pod whales—a subpod headed by L-22, Spirit, and her two sons: L-79, Skana, and L-89, Solstice—that hung out in the waters near Kanaka Bay for all that period of time, but they were actually quite an oddity, because they simply lingered in the same spot for the better part of three weeks, while no other whales were around. Resident orcas never do that; they almost always travel overnight and rarely ever stay in one place like that.

I paddled down there one day to observe them, and they were vocalizing a little, but they seemed to be hunting quite intently, because they were echolocating heavily. The whales all looked and sounded healthy, but with orcas, it is usually hard to tell if something is wrong until it's too late.

Bob Otis, a retired Ripon College professor from Wisconsin, monitors the Southern Residents closely every summer, along with a team of volunteers, from the perfect locale: the scenic lighthouse at Lime Kiln State Park, a place famous for whale sightings (indeed, the orcas come right up to the rocks there and play in the kelp and also seem fond of forming superpods in the waters in front of the prominent overlook, breaching and playing for the benefit of tourists). Otis keeps track of each day's sightings on a whiteboard calendar that can be viewed by the public so that people can know how many pods have been past each day. By the time he called it a wrap in mid-August in 2013, he had a board full of zeroes. Otis was shaking his head: "I've never seen anything like it."

Whale-watching boats, among the most popular tourist activities for the San Juan Islands, Vancouver, and Victoria, were relegated to either spending their time with the lonely three L-pod whales, or they went looking for other wildlife. As it happened, it was a good year in the Salish Sea for minke whales and humpbacks, who turned up in large numbers. There were other oddities that summer: a seventeen-foot, deep-diving sixgill shark, a fish no one in those parts had ever seen, washed up at Argyle Bay in Friday Harbor; the scientists who performed the necropsy found an empty stomach but no apparent cause of death.

No one is ever really certain about orca behavior, but the prevailing theory among the islands' whale scientists about their seeming disappearance was that the numbers of Chinook were probably down. The big king salmon comprise over 80 percent of the orcas' summer diet, and over 90 percent of the Southern Residents' summer Chinook typically come from the run that is heading up to the Fraser River in British Columbia, the mouth of which is just south of Vancouver. Sure enough, it soon emerged, as the fish counts arrived from the Canadian biologists who monitor these things, that the Fraser Chinook run was down in 2013 (along with the far more endangered sockeye runs in the river, victims of rising river temperatures).

However, it wasn't a terrible salmon run, just mediocre. Pretty soon it became clear that it just couldn't compete, because out on the continental shelf, west of Vancouver Island and the Washington and Oregon coasts, an incredible run of Chinook from the Columbia River—over a million fish, as it soon turned out—were gathering.

If there is anything we have learned about orcas from observing them over the years, it is that they go where the salmon are. Especially the Chinook salmon. A bounty that large, food that readily available, has become rare enough in recent years that it only made sense that they would descend upon it intently, remaining out on the Shelf until the salmon started to actually run up the Columbia. Indeed, when the orcas finally returned to solid Chinook runs in the Salish Sea in mid-August, they looked healthy and well-fed. So the conclusion was that the huge Columbia run had kept the orcas out of the islands for much of the summer. At least, that was the prevailing theory. Or maybe it was just the orcas being orcas: wily and unpredictable.

It also could have been boat noise; the numbers of whale-watching and private boats have been soaring in recent years, making for near-constant flotillas around the orcas during daytime appearances in the San Juan Islands. There has also been a steady increase in shipping traffic through Haro Strait, with more to come, adding to the overall roar of noise they must endure while hunting. Some have theorized that maybe the whales had just gotten tired of the fuss and decided to go eat where it would be nice and quiet.

Both of these factors—salmon plenitude and vessel traffic—are, as it happens, considered the two most important components of population recovery for these 81 endangered Southern Resident orcas, who were listed under the federal Endangered Species Act in 2005. The abundance of salmon is an obvious issue; if the orcas aren't getting enough to eat, they won't survive long. Vessel traffic is more complex, posing a kind of multi-level threat. The noise they create interferes with intrapod communication, which is critical for hunting, as well as with their ability to echolocate. The simple presence of all vessels, from kayaks to giant freighters and everything in between, alters the course that orcas travel and thus affects their ability to hunt and probably requires them to expend energy needlessly. Some of these vessels also carry real pollution risks. Pollution itself is considered the third major threat to recovering the Southern Resident population and getting them off the endangered list. It takes many forms, the most pernicious being the heavy metals and other toxins that accrue in the fish the whales eat and then in the whales' blubber.

Around the world, even among populations that are not endangered, a number of similar threats persist. In the much less densely human-populated home waters of British Columbia, the Northern Residents face similar threats, although on a much lower level, reflected in the considerably healthier population numbers they have been demonstrating for the past twenty years. A far more salient threat to this population has been the proliferation in their waters in the past couple of decades of salmon farming operations, an industry that brings pollution and predation problems with it, threatening a number of native salmon populations on which the orcas depend for their diet.

In the Crozet Islands, the population is threatened by its own tendency to steal fish from the longlines of fishermen who frequent their waters; these fishermen often kill orcas in retaliation. In Alaska, the resident killer whale population of Prince William Sound is still slowly recovering from the effects of the *Exxon Valdez* oil spill in 1989, and one transient pod, the AT1s, is now believed to be doomed, having lost over half of its members as a result of exposure to the spill.

However, the Southern Resident orcas of the Salish Sea are the only officially endangered orca population in the world. There has been little doubt for some time that, while boats and pollution are important contributors to the problem, far and away the biggest threat to their well-being has been the decline of salmon runs in the Pacific Northwest—not just in their home waters, but throughout the region, particularly on the Columbia River, whose massive runs of salmon historically were a major source of food for these orcas, notably during the winter and spring months, when Chinook runs are in decline inshore and (usually) abundant offshore.

So the massive Columbia River run of Chinook in 2013 was the best possible news for the whales' long-term recovery, even if one of its consequences was a reduced presence of the whales in the Salish Sea. In the end, everyone knows that saving the orcas means, first and above all, saving the salmon. What matters is that the whales stay well fed and, consequently, healthy.

"It sucks that we aren't able to see them more," observed long-time orca watcher Monika Wieland at her blog, "but really what's most important is that they're finding enough to eat."

Salmon biologists have credited the recovery on cooling waters and abundant food for the salmon in the Pacific Ocean for the past four years, and perhaps most significantly, a string of court-ordered improvements in salmon-recovery efforts throughout the Columbia River system that began over a decade ago and were finally proving to be effective. Biologists counted a million fish up the river by December. It was the largest Chinook run on official record, ever since the Bonneville Dam was built on the Columbia in 1938 (when they began keeping track), an event that inalterably transformed the river's ecosystem. Even for all that improvement, the run was still a shadow of the historic salmon abundance before the arrival of whites in the 19th century, when runs of salmon ran into the tens of millions. That was true not merely of the Columbia, but all of the Pacific Northwest, including the river basins throughout the Puget Sound/Salish Sea region.

It was the first positive sign of the real effectiveness of decades' worth of conservation efforts, an uptick that could mean a shift in the long downward trend of human effects on the natural environment of the Pacific Northwest, not just on killer whales and salmon, but on all the creatures tied up in the biological web that is the region's ecosystem. Or it might just have been a burp. What everyone involved understood was that, good news notwithstanding, the ecosystem's problems, embodied in the threats to its numerous endangered species, will not be solved overnight and that is because they have been a long time in the making.

• • •

The spirit of wanton extermination is rife; and it has been well remarked, it really seems as though the man would be loudly applauded who was discovered to have killed the last salmon.
—LIEUTENANT CAMPBELL HARDY, British Royal Artillery, commenting on wanton wildlife management in New Brunswick, 1855

Americans tend to think of wild salmon now as a West Coast phenomenon—reasonably, since there are no remaining wild runs of Atlantic salmon on the East Coast or in Europe and very little memory of them. In fact, salmon was once as abundant in Europe as it is today in Alaska. The sad history of what became of Europe's salmon was replicated on the East Coast of the Americas as well, and much of the fight over Pacific salmon today is part of a desperate attempt to not repeat that history one final and fatal time.

Salmon are a uniquely useful species for vibrant and healthy forest ecosystems. As young hatchlings and fry, they consume relatively little in the way of food resources in the waters of their birth or as they travel downstream to the ocean. Then, after spending years in the ocean, effectively gathering up proteins and becoming themselves large swimming bags of amino acids and other nutrients from the sea, they return with all those nutrients to the upper reaches of the rivers whence they came, where they die, bringing all those fresh proteins to regions where they become essential contributions to the food supply and the ecosystem.

In Europe, Atlantic salmon were historically abundant for most of recorded history, until the early 1700s, when human populations overfished, blocked rivers, and began destroying habitat for the many hundreds of salmon runs in France, Germany, the British Isles, Belgium, the Netherlands, and Scandinavia. By the 1800s, many runs were extinct, and fresh salmon became rare and expensive. Today, most European salmon runs are completely extinct, except for a handful of runs in the United Kingdom.

This story repeated itself when Europeans came to North America and began exploiting with wanton abandon both the salmon runs and the rivers in which the salmon swam. Indeed, as Lieutenant Hardy trenchantly observed at the time, the whole ethos of the settlers seemed at times to have been one of competitive eliminationism—the complete extirpation of anything deemed undesirable or dangerous or wild—all in the name of "conquering the wilderness." This ethos was consigned not merely to salmon but to nearly every indigenous occupant of the landscape, including the Native American occupants, who were frequently viewed and treated as subhuman by the new arrivals.

The results were predictable: By 1896, salmon had been wiped out of Lake Ontario and soon most of the remaining Great Lakes. Hundreds of other runs became extinct after the rivers were dammed. By the 1950s, most of the remaining Atlantic salmon runs originated from rivers in Maine, New Brunswick, and Nova Scotia, but those soon crashed altogether when fishing operations began netting large numbers of salmon in the open sea of the North Atlantic.

Almost as soon as Europeans began settling on the West Coast in large numbers in the 1850s, that story began repeating itself once again. However, the salmon there were so abundant that it has taken longer for the runs to be destroyed altogether, which has given people enough time to reconsider the path they were heading down. In the Northwest particularly, salmon are deeply tied to the identity of the region, including those deep forest regions where the salmon used to make their runs. There is also a deep biological connection; scientists estimate that 137 other species of animals in the Northwest depend on the presence of salmon.

In central Idaho, in the upper reaches of the Salmon River, which eventually drains into the Columbia via the Snake, settlers in the mining district of the Yankee Fork used to see such immense salmon runs during spawning season that the banks of the river overflowed with reeking dead salmon, creating a stench that lasted for weeks. I can still remember seeing, as a child in the early 1960s, sockeye runs on these same headwaters. At the spawning grounds, the water boiled with thrashing red, huge, scarred fish with vicious-looking hook jaws, and they were so thick I thought I might have been able walk across them.

That was nothing, of course, compared to the runs that used to be seen on the main Columbia, the largest river on the West Coast. When Native Americans dwelled along the river, they would set up clever fish traps called seines and would feed their tribes throughout the year on the bounties they could easily gather there. With the arrival of Europeans in the 1870s, however, the seines became massive river-wide operations that caught every fish that came through, all for the benefit of canneries that set up operations next to the seines. At first, their catches were phenomenally large, drawing from runs that were estimated to be 16 million and more. However, the takes at the seines were so complete, and the wastefulness of the operations so wanton and common, that by the early 1900s, the runs were in serious decline, and in 1908, Oregon voters passed the first initiatives to limit salmon fishing on the Columbia River. Eventually, the seines were banned altogether.

The next blow to the salmon came in the form of the dams that, starting in 1938, began going up on the Columbia. The construction of the Bonneville Dam, a popular WPA project that brought electricity to much of the rural Northwest, had salmon ladders that migrating fish could travel through. In 1942, the Grand Coulee Dam in central Washington was completed, and it had no fish passages at all. That meant the extinction of all salmon runs from rivers north of it in Canada.

The runs up the Snake and Salmon rivers, which join the Columbia below Grand Coulee, remained more or less intact. However, over the ensuing years, three more dams were built on the Columbia along the route to the Snake, and four were built on the lower Snake. It was these latter that especially spelled the death knell for many of the river's remaining salmon runs.

The four dams in question (Ice Harbor, Lower Monumental, Little Goose, and Lower Granite) were built starting in the 1960s and are located on the Snake River between Washington's Tri-Cities and Lewiston, Idaho. They provide irrigation water for a handful of large farms along the river, but their far larger role is to provide navigation for barges that provide cheap transportation of goods downriver for the region's farmers. The dams also generate 1,250 megawatts of power annually, enough for a city the size of Seattle.

However, they also drive salmon runs to the brink of extinction. The dams themselves are obstacle enough for migrating fish; the young smolt going downriver are chewed up in their turbines, and the returning spawners have to climb past them. In addition, the flatwater created behind the dams is even more problematic, especially for the smolt, who require fast-moving water to migrate effectively.

By the time I was a teenager in the 1970s, those massive salmon runs had almost vanished; there were only a handful. By 1992, only a single sockeye salmon, dubbed Lonesome Larry, managed to make it all the way to the Salmon River headwaters near Stanley Lake. In 1993, the government listed the entire Snake River sockeye run under the Endangered Species Act. In all, over a hundred Pacific salmon runs in the Pacific Northwest have gone extinct; another 200, representing over half of the total population, are considered "depressed" and at risk of extinction.

Spilling water over the dams' gates at key times of year, so that the smolt can run downstream freely en route to the ocean, has proved effective as a remedial step, but that also conflicts with the dams' power-generating mission. The solution of the Bonneville Power Administration, the lead government agency involved in the Columbia River salmon recovery, has been to gather the smolt as they head downriver and barge them around the dams. This is an exorbitantly expensive program that also has proved to be of questionable real value, since the runs have been slow to recover.

Salmon advocates like Save Our Wild Salmon have argued for tearing down the dams as the most sensible solution, since it would return that portion of the river to a free-flowing state and give both smolt and spawners a fighting chance of success. They argue that the economic costs can be overcome, pointing out that replacing the barges with a revamped rail transportation system and simply lowering the current irrigation pumps would cost a fraction of the current barging system. Moreover, they point to larger economic benefits for the region, particularly the economic boon that could be realized from recreation.

These arguments, however, have carried little weight with eastern Washingtonians, who have come to see the dams as emblematic of "their way of life" and thus to be defended at all costs. When the breaching was first proposed in 1999, pro-dam rallies were held in various communities at which the rhetoric became high-pitched. Leading the way were top Republican officials, including then-Senator Slade Gorton, who warned of various miseries that breaching would inflict.

"We are not going to allow a few Seattle ultraliberal environmental zealots to destroy what took generations to build," said state Senator Dan McDonald, R-Bellevue, at a Richland gathering.

"In case you don't understand the urgency of this, think about this: The bulldozers are coming," said Representative Shirley Hankins, R-Richland. "The gun is at our heads, and we need to act right now before they pull the trigger."

Since 2001, the federal government (first, the Bush administration, followed by a compliant Obama administration) has officially opposed any breaching program, reverting to a reliance on barging. A federal judge's May 2005 ruling that the barging program is failing and demand that the government reexamine its salmon-recovery progress was greeted with warnings from the Dry Side that doing so had better not put dam removal back on the table: "Changes may need to be made, but the dams are going nowhere," said Representative Doc Hastings, R-Wash.

In 2006, an accord was announced that required water to be spilled at critical times to aid salmon runs, while compensating farmers and finding ways to prepare for their water needs during those times. This tamped down the conflict by bolstering salmon runs semi-sufficiently without dam removal, and if the results six years (the typical life cycle of the salmon) after the accord took effect are anything to judge by, the plan is nothing short of a celebration-worthy success. A million Chinook are clear testament to what can happen if even modest conservation efforts are pursued and effectively maintained.

Still, salmon scientists and wild-river advocates have been adamant all along that the only way to bring back the salmon runs to a modicum of health will require a free-flowing Snake River, something spills don't achieve; to that end, they have continued to advocate tearing down those dams.

• • •

The listing of the resident killer whales under the Endangered Species Act (ESA) in 2006 could prove critical in this debate. If it is established that the Southern Residents, in fact, depend on Columbia River Chinook salmon, then their needs will certainly play a role in how the policies eventually are adjudicated.

The decision to list the orcas (currently about 80 whales) was announced late in 2005 after years of pressure from environmental and whale-advocacy groups and was officially put in place in the spring of 2006 by the National Marine Fisheries Service (NMFS). The announcement was generally greeted with fulsome praise in the local press and among civic and chamber-of-commerce leaders. After all, not only are the whales now a multi-million-dollar ecotourist attraction, but they are above all else a regional icon, part of the Northwest identity. However, salmon-recovery efforts in Puget Sound, notably on the Skagit and Nisqually rivers and the Lake Washington watershed, have been slow to produce salmon numbers. There has been improvement, but it has been mostly incremental. The whales have been scraping on their resources in their home waters for some time now and still are.

Killer whales are prodigious eaters; their daily caloric requirements are about 10 percent of their body weight, and adults range in size between 8,000 and 15,000 pounds. That means a single Southern Resident will eat as much as 1,500 pounds of salmon in a day. Probably the best public demonstration of their voraciousness and their increasing desperation occurred in 1997 at Dyes Inlet, a remote cove deep in southern Puget Sound, when a contingent of 19 L Pod orcas hung out at the mouth of the inlet for two months and consumed nearly the entirety of a substantial run of chum salmon. That event, notable not least because L Pod rarely strays that far south in the Sound and usually has a strong preference for Chinook salmon, appeared to signal that the Southern Residents were having real trouble finding enough food to eat. Over the next five years, the population of Southern Residents began to decline precipitously, from nearly 100 in the early 1990s to 79 by 2001.

Scientists grew concerned that if the declines continued much further, the whales would no longer have a viable gene pool and would soon tumble into an inevitable downward population spiral. By then, scientists had learned that the Southern Residents are genetically isolated, that even though Bigg's killer whales visit their waters to feed on large sea mammals like seals and sea lions, there is no social interaction or apparent communication between them. Likewise, they do not interact with their fellow Northern Residents at all, nor with offshore killer whales, as far as we know. So the presence of a substantial coastal population of transient and offshore orcas means little to the killer whales who reside in Puget Sound for much of the year. Recognizing that this was indeed a distinct population proved key in NMFS's decision to list the orcas as endangered, since their previous approach had been to consider them simply a subset of the larger coastal population.

"We were petitioned to list back in 2001," recalls the Northwest's chief whale scientist for NMFS, Brad Hanson, "and we went through the status review process for distinct population segments. There was a lot of angst in the process because the whole question boiled down to taxonomy, and at the time, killer whales were considered one species worldwide. So when you look at the factors, the point was, because transients and residents overlapped in the range, if you lost Southern Residents, would that cause a major gap in the population range? Well, the answer is no, it wouldn't, because you'd still have transients. Really, there wasn't enough data at that point. The bottom line was that nobody had even suggested then that there might be separate species or subspecies. So we said no at that point."

Three years and a citizen lawsuit later, NOAA officials went back to re-examine the matter. "In 2004, there was a cetacean taxonomy workshop where we had all the people working on killer whale genetics in one room," recalls Hanson. "There was a general feeling that there was enough evidence to suggest that it might be a subspecies. We included a taxonomist in our work group and ultimately concluded there was enough evidence to warrant considering it a subspecies. So that qualified it for a distinct population segment."

Once that was decided, it became almost an inevitability that the Southern Residents would be listed, and the taxonomic decision was eventually confirmed in 2010, when NOAA scientist Philip Morin published his study of mitochondrial DNA in resident and Bigg's orcas, demonstrating that there had been no genetic crossover between the two populations for over 700,000 years.

The sharpness of the 1990s declines also made plain just how vulnerable to extinction the Southern Residents had become. Those declines, almost certainly not coincidentally, occurred simultaneously with large drops in salmon stocks. So, while the NMFS recovery program recognizes that the whales actually face a variety of threats to their well-being, it is also clear that the cornerstone of any recovery for the orcas lies with producing enough salmon for them to eat.

The larger problem of salmon abundance, moreover, is intertwined with subsidiary issues such as pollution and vessel traffic. Noise in the water and the presence of boats is mostly a problem for orcas when they are already stressed searching for scarce salmon; it just makes their work that much harder. More importantly, when orcas run out of food, their bodies begin burning their fat reserves, and those stored toxins begin circulating in their systems and poisoning them. So the ESA listing of the Southern Residents meant that many of the efforts currently under way on behalf of the fish, including a $1 billion Puget Sound salmon recovery plan begun at the same time, would have even more teeth, so to speak. Many of these plans include placing restrictions on development in sensitive areas and limiting runoffs from urban areas into the orcas' waters. Moreover, NMFS's orca-recovery plan specifically required "expansion of local land-use planning and

control, including management of future growth and development."
That immediately affected developers and the construction industry,
as well as various property owners, whose lands are in areas that either
affect salmon runs or orca habitat, which is much of the Sound.

Some of the listing's other effects included:

- Examining the adequacy of wastewater-treatment plants in the region and perhaps requiring upgrades.
- Confronting the wastes being dumped into the Sound by cruise ships.
- Potentially regulating whale-watching operations.
- Enhancing cleanup efforts at toxic-waste sites, particularly those containing polychlorinated biphenyls or PCBs.

Unsurprisingly, development and construction interests immediately weighed in; in early 2006, the Building Industry Association of Washington (BIAW), along with the Washington Farm Bureau Federation, filed a lawsuit contesting the listing. BIAW attorney Tom Harris told a Seattle reporter that it was "an unlawful listing," adding, "You can almost say any individual school of fish can be listed." A federal judge dismissed the lawsuit several months later.

The legal beagles who cobbled that lawsuit together, the conservative Pacific Legal Foundation, attempted more or less the same lawsuit in 2012 on behalf of California property owners involved in the water-rights battles along the Sacramento River, claiming that the orcas' endangered status was about to rob them of their water rights. A judge dismissed that lawsuit, too.

There may be other bones of contention with powerful interests yet to come. American wildlife managers may raise concerns over management of the Fraser River fishery, which is a significant food source for the Southern Residents. However, the Fraser is not the only major river outside of the Puget Sound whose runs are part of the orcas' food picture. Another, in fact, is the Columbia River, and that puts those dams eventually in the crosshairs.

• • •

The problem scientists face right now when it comes to assessing how to proceed with the orcas is a real lack of data. They're unsure which salmon the orcas are eating at which times of year and which runs are truly critical for their well-being. Underlying the uncertainty is one of the orcas' abiding mysteries, namely, where they go and what they eat during the winter months.

What we know about the orcas is mostly based on what we observe of them when they're in Puget Sound and, for the majority of them, that means the months of May through September. J Pod will continue lurking about in the Sound throughout the year, although even that is sporadic; they're often observed as far south as Vashon Island in the winter months. However, the K and L pods, which constitute the large majority of the clan, head offshore, but no one is sure exactly where they go, since they spread out and are only sporadically sighted.

The first scientist to try to tackle these mysteries was John Ford, the renowned orca researcher and author and a biologist for the Canada Department of Fisheries and Oceans. In his 2006 study ("Linking Prey and Population Dynamics: Did Food Limitation Cause Recent Declines of 'Resident' Killer Whales in British Columbia?"), he found, among other things, that the Southern Residents range, in the winter months, from as far north as the Queen Charlotte Islands (where L pod has been sighted, and a K pod member washed ashore dead) to as far south as northern California, where the J and K pods have been sighted. However, they stay relatively close to shore; none have been observed any farther from the coast than about 30 miles.

His core finding, however, was linking the whales with a particular kind of salmon: Chinook. Orcas focus on them almost exclusively in the summer months, Ford found: "Chinook salmon appears to be preferred over other salmonid and nonsalmonid species due to its large body size, high lipid content, and year-round availability in the whales' coastal habitat. Sockeye and pink salmon, which are abundant during migrations to spawning rivers in July-August, are not a significant prey species." Some runs of chum salmon appear to be preferred in the fall.

Moreover, Ford notes, "The distribution and movement patterns of resident killer whales are consistent with what might be expected of an animal having a year-round focus on Chinook salmon as preferred prey." That is, during those winter months, they are haunting waters that are historically known to contain large runs of Chinook, gathering along the coastlines on their way home to their respective rivers to spawn.

"I think there's been a lot of assumptions over the years, and they're very reasonable ones," Ford told me, "that the whales feed from the spectrum of salmonid species that are available to them, especially the most abundant ones. I guess that's what's really changed as a result of our work, that we've now convinced ourselves that in fact they're very selective. It's really Chinook that seems to be of critical concern to the whales and us."

Historically, the largest single source of Chinook in the North-west's Pacific coastal waters during the winter and spring has been one place: the Columbia River. The role that the Chinook salmon could play in the orcas' health was underscored two years ago by a Washington Department of Fish and Wildlife report on killer whales: "Perhaps the single greatest change in food availability for resident killer whales since the late 1800s has been the decline of salmon in the Columbia River basin."

So far, the lack of hard data keeps scientists from concretely link-ing the Southern Residents with the Columbia River Chinook, but Ford says the data uncovered in his study tends to point in that direc-tion: "Part of our study is that we actually genotyped all the Chinook samples we got from the whales, and we're putting a piece together now about which river systems the whales are taking," he said in the interview. "For example, we've got a fair number of Chinook samples from Northern Residents up in the northern end of the Queen Char-lotte Islands, and a substantial proportion of the salmon that the whales take up there are Columbia River fish on their southward migration."

Ford says that the ecosystem is large enough to involve a broad range of river systems: "No doubt there's a lot of Chinook populations that are of questionable status, and that's the next step, to see how they can be better conserved.

"Because killer whales don't have any predators, they are ultimately prey-limited. The question is, are they at that sort of carrying capacity? And when that carrying capacity declines unexpectedly, as it did in the late '90s, do they suffer? It makes sense, therefore, that enhancing the availability of Chinook for the whales, sort of both in quality and quantity and temporal or seasonal availability would be a pretty reasonable recovery strategy."

What that involves, however, is another question. "We don't have a complete enough understanding of the whales' diet, especially in the winter," says Ford. "And that's sort of what we need next to understand."

Whale advocates say they've been aware of the potential connection of the orcas to the Columbia River runs for a while: "This is something I've been talking about for a long time," says Howard Garrett of the Whidbey Island-based Orca Network. "We've known it almost intuitively. It's been part of my regular slide show. So, it's gratifying to see the scientific data supporting it."

"To me, it's just a no-brainer," says Darcie Larson, a board member of the Seattle chapter of the American Cetacean Society and the associate director of Save Our Wild Salmon. "It couldn't be more plain that these killer whales have relied on salmon from the Columbia River historically, and the lack of those salmon is hurting them now.

"I think even without definitive studies, you can still clearly say that, before the time of Lewis and Clark, there were between 10 and 16 million wild fish that were returning to the mouth of the Columbia River. We know that whales are eating these fish, we know that they like Chinook, and there were definitely millions of Chinook that were returning to the mouth of the Columbia."

There is also historical anecdotal evidence that puts killer whales at the mouth of the Columbia during the winter and spring months. More recently, a group of J Pod orcas was observed feeding on salmon there. However, these remain anecdotes and intuitions. In the winter of 2012, they finally began obtaining hard evidence of it: NOAA scientists successfully attached a sensor to a K pod whale and were able to track the pod's movement's for several months, during a time when the orcas are difficult to study. What they found was that, not only did the pod travel all the way down the coast to Monterey, but they spent an inordinate amount of time hanging out at the mouth of the Columbia. When the Columbia River fish enter the equation for orcas, their endangered-species status takes on even larger political ramifications, because that population of Chinook is directly affected by those same four dams on the Snake River that have become a major battlefield in the state.

Killer whales, after all, reside atop Puget Sound's food chain and are thus one of the real indicator species for the overall health of its inland waters. If they disappear, it will toll a death knell for much more than just whales. The bigger picture, as Brent Norberg suggests, is that the orca listing is already certain to have a positive impact on the Puget Sound ecosystem, and perhaps beyond it as well, by underscoring that they are simply some of the most prominent occupants of a vast and complex ecosystem.

"I think in general what gets missed in the public mind is that there are substantial things being done already on the part of fish and clean water and so on before you ever get to the whale link," Norberg said. "The whales, however, because of their charismatic position in the public consciousness, make a really good focal point to try and leverage more beneficial actions on the part of those other things that are already being done. They make a positive argument for doing things to benefit fish. They help make a positive argument for doing things to clean up persistent pollution of sediments in the water. Those are good things that are going to benefit not only the whales in the long term, but other species, too."

The defenders of the dams, in the meantime, have insisted all along that there's no evidence that tearing them down will even restore the salmon to their runs anyway and that proposals to tear down the dams are nothing but pie-in-the-sky environmentalist fantasies. However, on the Elwha River, that assertion is being put to the test.

• • •

No matter how far over the edge you dare to peek, it is so deep and the space so dark and narrow that you can't really see the bottom of the carved-stone gap through which the Elwha River is pouring at the top of Glines Canyon. The water roars over the top of the dam down into the canyon, but forget about being able to see the river 210 feet below. All you can see are the mists that float up into the canyon and turn the black rock walls sleek and mossy.

The mouth of the waterfall is lower now; at the sides of the gap, you can see the marks where the dam once was, now blasted away, bit by bit. At one time the dam reached up to the level where most people can now stand, looking down into the gap. Now it is down to the point where the waterfall is, some thirty feet below, and getting lower with each blast of dynamite to bring it down. Behind the dam, or at least its remnants, there is a vast mud flat stretching back several miles where there used to be a reservoir called Lake Mills. Running down through the middle of it is the now-recovering Elwha River. This is all part of a grand experiment, the purpose of which is to find out if salmon runs can return to rivers if the dams that made them extinct are removed. Even though some 500 small dams have been demolished in the United States in the past twenty years, something of this scale has never been attempted before.

The Glines Canyon Dam is 13 miles away from the Strait of Juan de Fuca. Eight miles downstream from it are the remains of the old Elwha Dam, now completely removed. The dams on the Elwha are among the earliest dams built in the Northwest. Elwha Dam, first built in 1912, was a hastily erected travesty with no fish ladders, made so badly that it collapsed the first time it filled, and then was rebuilt in 1913, all in defiance of the state's laws governing such structures. Entrepreneurial city fathers from nearby Port Angeles, intent on having hydroelectric power to run the timber mills that they intended as the backbone of their local economy, were behind its construction, as well as that of Glines Canyon Dam in 1927. The latter was actually quite an engineering feat, involving not just building a concrete structure in a narrow, steep canyon, but also drilling tubes down through the rock-canyon face through which water would pour to push the dam's turbines. Of course, it had no passages for salmon, either.

At one time, the Elwha River had massive salmon runs that ran deep into the heart of what is now Olympic National Park, a wild and pristine place, rich with vegetation and wildlife. It was one of the only rivers that contained stocks of all five species of Pacific salmon and was considered the most bountiful source of fish, particularly its huge Chinook, on the Olympic Peninsula. The Elwha dams almost completely eradicated these runs within a few years of their construction. By the 1970s, however, the local timber mills in Port Angeles and around the Peninsula were part of the large Northwest power grid and no longer dependent on the meager output of the old Elwha dams. At that point, people began to ask: Are those dams worth the cost?

First to raise the issue were the Lower Elwha Klallam tribe, who contested relicensing of the dams before the Federal Energy Regulatory Committee beginning in 1973 and were soon joined in the effort by a variety of environmental groups. By the 1980s, a broad coalition of environmental and tribal groups had coalesced around the idea of tearing down the Elwha Dams and restoring the river. The efforts resulted in the 1992 Elwha River Ecosystem and Fisheries Restoration Act, passed by Congress and signed into law, but then effectively blocked and delayed by Washington's Republican U.S. Senator at the time, Slade Gorton. Finally, in 2000, Gorton relented, and the government went about the work of purchasing the dams and preparing for restoration work on the river. They also began drawing down the reservoirs behind the dams so they would be mostly empty when the time came.

In September 2011, construction crews began tearing down Elwha Dam and were done in spring of 2012, well ahead of schedule. The Glines Canyon Dam teardown, begun shortly thereafter, has taken longer, mainly because of the amount of sediment behind it; officials underestimated just how much sediment would be making its way downstream and soon found the systems they had built to deal with it clogged. So they delayed the finishing touches on the Glines Canyon Dam to allow the systems more time to deal with the combination of fine grit and heavy sand that was pouring downstream. The last fifty feet of the dam remained in place for most of 2013 while restoration overseers waited. It finally came down completely in the summer of 2014.

The Elwha project has ramifications well outside of the cloistered confines of the Northwest; if it proves successful, it could weigh heavily in the debates over a number of dams throughout the United States. Most of the 84,000 dams in America are aging—some 70 percent of them will be older than 50 in the year 2020—and will increasingly need repair, replacement, or removal. Some of them need it immediately; a substantial number (4,400) are crumbling and considered serious failure risks, and repairs will not be cheap, costing upwards of $21 billion.

The Elwha is considered one of the most ambitious ecosystem-restoration projects ever, and even with all the problems the project has experienced, the early results have already been remarkable. In the fall of 2013, biologists went out in snorkel gear and wet suits to count salmon returns in the Elwha below Glines. What they found surprised them; nearly two thousand Chinook and seven hundred new Chinook spawning nests, representing a potential bounty of even thousands more adult fish returning in another six or seven years. "It is truly exciting to see the Chinook finding their way into clear water tributaries," remarked the parks superintendent. "This is what we have always known was coming."

The good news from the Elwha was also good news for orca advocates. "I'm very eager to see what happens if they can restore the Chinook run in the Elwha," says Ken Balcomb. "It won't solve everything, but every little bit of improvement helps the whales."

Indeed, whale scientists have been quick to warn orca advocates not to get too excited about the Elwha restoration as a solution for the Southern Residents. "A lot of people get really excited about the dam removal on the Elwha, and don't get me wrong, it's great," says Brad Hanson. "But I think there are some people who think that's gonna save killer whales. And the truth is, it's barely a pimple on a bear's butt. It's incremental. People don't understand the number of fish that need to be in the system. You think about what the Fraser River produces; we're talking about hundreds of thousands of fish that need to be produced. So yeah, it will help, every bit helps, but it's just a little bit of where we need to get."

Hanson recommends looking elsewhere for a real impact that would help orcas and salmon stocks generally: "I think the best bang for the buck is to go into systems like the Skagit and the Nisqually, those bigger river systems that are still in pretty good shape, and try to make sure that those are enhanced. And that's a challenge."

Killer whales, for that matter, have not been seen much along the shores of the Olympic Peninsula for decades, at least not since the Elwha Chinook runs dwindled to nothing. However, their prevalence in the mythology of the Klallam tribesmen who occupied the area suggests that one time they were common there. The return of Chinook to the Elwha has raised a lot of people's hopes that the killer whales, too, might return there to feed.

"That would be a great thing to see," says Howard Garrett. "If nothing else, it would mean they had yet another place to get their salmon."

• • •

One orca came back, although not in a good way. The body of a female Bigg's killer whale washed up on the shores of the Olympic Peninsula in early January 2002. She had come a long way to die: Identified as CA189 and named Hope by Northern California whale watchers, a transient who normally occupied the waters off the northern California coast, she was found washed up in a marshy area near Dungeness Spit. Not far from her body was a young male orca, very much alive, soon ID'd as her son, CA188. He had nearly beached himself, but after some effort, he was eventually guided into deeper water, where he swam away to parts unknown.

It took some work to bring the body of Hope into shore, and once there, scientists immediately began performing a necropsy and were stunned and alarmed by what they found in the orca's outer layers of fat. Scanning for the toxin known as polychlorinated biphenyls, better known as PCBs, they found levels that were literally off the charts, so high that they had to recalibrate their instruments to measure it: 1,000 parts of PCB per million parts of whale fat. She also had abnormally high levels of other toxins, some of which were completely unexpected. "She basically knocked our instruments off," said NMFS researcher Gina Ylitalo. "We had no idea we'd see these levels." Everyone working on Hope was required to wear a hazmat suit, and the site around her body was treated as a hazardous-waste disposal operation.

None of this was exactly a surprise to Dr. Peter Ross, a toxicologist with Canada's Department of Fisheries and Oceans, who has been amassing evidence regarding toxins in marine mammals for years. As Ross puts it: "Killer whales rank among the world's most contaminated marine mammals."

The problem arises from their position atop the oceanic food chain. Long-lived toxins, called "persistent organic pollutants" (or POPs), remain in the food chain because they are stored in the fatty deposits of various animals, from fish all the way up to seals and, ultimately, whales. They harm orca populations in two main ways: first, through the milk provided by female orcas, whose fat stores provide much of the source of the rich milk they feed to infant orcas. When these toxins are passed directly to the young whales, they poison many of them and are usually blamed for the high infant mortality rate of wild orcas. They are also believed to play a role in the mortality of adult orcas such as Hope, whose death ultimately was assigned to a complex of problems arising from all the toxins in her body, and others, both Bigg's and resident orcas, when they are stressed for food sources. At those times, their bodies begin processing their fat stores, and the toxins are released into their systems. The toxins aren't believed to kill the orcas outright, but rather, to compromise their immune systems so badly that they become susceptible to a variety of diseases, especially lung infections.

There are three specific kinds of POPs that affect killer whales:

- **PCBs.** These long-lived pollutants originated at the turn of the last century and were used for most of the next seventy years in paints, electrical transformers, sealants, adhesives, and hydraulic fluids, and are found commonly in Puget Sound. Because the molecules are so long-lived, they are still leaching into watersheds from various dump sites around the Sound, notably at the various military bases that dot its waterfronts. PCBs can impair reproduction and growth in mammals and are blamed for making them susceptible to infections.

- **Polybrominated diphenyl ethers, or PDBEs.** These are relatively recent arrivals on the toxin scene, but they are ubiquitous now, used as fire retardants in a large number of consumer products, including furniture and computer products. Their effect on mammals is similar to that of PCBs. They make their way into the aquatic food chain in large part by washing out to sea with runoff from urban areas.
- **Dioxins and furans.** These closely related chemicals are poisons produced by burning organic material in the presence of chlorine and were for years associated with the papermaking process employed by the pulp mills common throughout the Pacific Northwest, including Canada. They also appear as a byproduct of the activity of municipal incinerators, coal-fired generators, diesel engines, and sewage sludge.

"Contaminants don't act as a rapid poison," says Ross, "but weaken the whales' ability to survive diseases or other stresses. In southern B.C. and northern Washington State, we've got about seven million people, as against 80-some killer whales in the local Southern Resident population. In sharing killer whale habitat, citizens of this region can act as partners with government and industry when they consider their responsibilities in light of household cleaners, pesticides, vehicle use and other lifestyle choices. By protecting streams, rivers and our coastal waters, we can help ensure the survival of salmon and killer whales into the future."

• • •

The J Pod has a new calf, and it seems to be showing off. It's not just frolicking with the grownup orcas like many of the young like to do;

Chris, the chief naturalist for the Western Prince, *points out distant whale signs for sightseers.*

it's breaching repeatedly, coming up sometimes among a whole cluster of older whales and leaping fully out of the water. It's an infectious display of delight in the pure joy of living that seems to affect all of us watching. We laugh, smile, applaud, ooh and aah. Stationed at the front of the boat, we're getting quite a display. There is only a handful of us, and I wonder why everyone else is staying at the back end of the boat.

"Whoa!" The whale watchers in the back holler in unison, followed by whoops and applause. Looking back, those of us in front can quickly see why: Another juvenile orca—this one a good deal larger than the newborn we've been watching—is breaching about a hundred yards off the back end of the boat, multiple times: once, twice, thrice, even a fourth time. With each successive leap, it seems determined to come higher out of the water, until on the last few breaches it completely clears the water, its tail tucked sideways like a salmon's.

With each leap, the cries from the small knot of people at the end of the *Western Prince*, a 30-foot cruiser that runs daily tours out of Friday

Harbor, grows louder. These are people from all over the United States, and indeed the world, who have signed up to see something like this, although none of them are really prepared for the experience. A tow-headed eight-year-old boy from Lynnwood is especially rapt: "Wow!" he keeps saying. "Wow!"

Not everyone who signs onto a whale-watching boat gets to see this (in fact, this is my third trip out with the *Western Prince*, but the first time I've seen the orcas with them), but for nearly every one of the 25 people on board, this is why they're out on the water in the San Juan Islands this sunny day in early May: to see killer whales.

There are only about 24 killer whales in J pod. Make that 25 with the new calf, although it does not count officially for about a year, after it has made it through the early stages that are proving extraordinarily hazardous for young orcas. During the course of a typical summer, over half a million people will board whale-watching boats based in Seattle, Vancouver, Victoria, Friday Harbor, and elsewhere in the San Juan Islands and the Puget Sound region. Many of them are people who arrive here from all over the world, hoping to get a glimpse of our famous orcas. It's big money. Industry officials and regional chambers of commerce say the whale-watching industry attracts several million dollars (the estimates vary) annually to the region, and the demand keeps growing.

Yet beneath all the sunny exclamations and beatific testimonials, there is a dark side to the whale-watching industry, directly connected to the doom that hangs over the whales themselves and all those people coming to see them. When the Southern Residents went on the endangered-species list, one of the major factors that federal officials began looking at in helping them recover was the potential need to regulate the whale-watching industry. The problem is self-evident to anyone who witnesses the scene off the western coast of San Juan Island on a typical summer's day. Whenever orcas arrive on the scene, they are accompanied by a massive flotilla of boats: whale-watching tour craft intermingled with private recreational boaters out for a gander, as well as fishermen and other regular users of these waters, including kayakers out looking for whales, too. They zoom in and out and around, creating a scene that often borders on the frenetic, all in pursuit of a close glimpse of the largest member of the dolphin family. It's constant; at times during the summer, whale observers say the orcas are accompanied by boats from nearly sunup to sundown.

It's even more apparent to anyone who drops a hydrophone beneath the surface of Haro Strait at these times to listen in on the whales' conversations, which come in the form of whistles, chirps, calls, and clicks. In addition to whatever noise the whales make, the strait itself is a cacophonic canyon full of boat-engine noise. The sounds range from the high squeals put up by small outboard motors zooming through to the sometimes overpowering thrumming noise created by the steady stream of large vessels en route through the strait to Vancouver and points north. Scientists have been studying the effects of all these boats and the noise they create on the whales in recent years, and the data they've collected so far indicates that, at least in years when the supply of Chinook salmon that comprise the bulk of their diet is low, the boats are amplifying the harm to the whales.

"There are two things of concern, really," says David Bain, a former University of Washington marine biologist who specializes in killer whales. "One is that they tend to travel farther to get from Point A to Point B when boats are around. So it's kind of like they take detours around the boats. So basically what that means is they're spending more energy than they would otherwise, because some of the whales are kind of zigzagging around the boats, and other whales go straight through, and they stop and rest while the others catch up. There's a difference in how much energy they spend swimming versus how much energy they spend while they're resting. The other effect we're seeing is they do less foraging when boats are around. That probably means that they're eating less and acquiring less energy. There is an energy balance in whales, and whale watching has an effect on that."

In some regards, this means that, for the whales, a critical mass of obstacles is created by the sheer number of vessels, including the kayaks, which Bain says are capable of startling the whales in ways that power boats cannot, especially if they dart into the whales' path or invade their space, and fail to warn the whales of their presence. Still, a kayak that observes the preferred whale-watching guideline of 100 yards' distance will have almost no effect on the whales because of its silence, while any power boat within audible range is creating at least some level of disturbance.

"Most likely noise is a mechanism involved," Bain says. "Killer whales find their food with echolocation, and the noise reduces the range at which they can detect prey with echolocation. So my sense of what's happening is that when it gets noisy, they don't even try to look for fish. They just kind of save their energy and hope they run into fish in a quiet place or a fish gets close enough that they can in essence bump into it or find it without really looking for it."

The noise levels, as well as the kinds of noise, vary widely from boat to boat. For the most part, the majority of whale-watching boats are not particularly noisy; there are just a lot of them. The loudest noise comes from shipping traffic, although much of that is lower-frequency noise that whales often seem to talk over. The most disruptive noise comes from private recreational boaters, particularly those who fire up their engines in close proximity to the whales.

"The smaller boats tend to make higher-frequency noise, so it interferes more directly with the killer whales' communication and echolocation signals," says Bain. "And the larger ships tend to make lower-frequency noise, which is less relevant to the killer whales. I think what the noise means is that they're more likely to miss the fish that are there. Because of that, it may not be worth the effort for them to, say, dive down to where the fish are most likely to be. They'll travel along and try to find a quiet spot and then they'll fish there, so they'll have in essence less time to find food."

There are several mostly volunteer efforts under way to get the chaos under some semblance of control. Probably the most visible of these is SoundWatch, a long-time Friday Harbor-based program operated under the auspices of the Whale Museum, which sends boats out to monitor the scene in the San Juan Islands in the summer, approaching boaters who violate the whale-watching guidelines and urging them to respect the rules.

SoundWatch and a local activist group, Orca Relief Citizens Alliance, have approached San Juan County officials about revisiting the possibility of regulating vessel traffic around the whales. "Now that we have this Endangered Species Act listing, there's actually some new languages and new provisions that really give states and counties a little bit more local authority to enact conservation measures, as long as they are on par with the goals of recovery plans within the ESA, to do everything from speed limits to jet-ski bans," Kari Koski of SoundWatch told me. "So the county is revisiting it, not really to say, 'Let's sock it to those commercial whale watchers,' but more to say, 'How do we level the playing field a little bit? How do we make it be more in the consciousness of boaters, every boater who's on the water, that they need to be paying attention and be accountable for their actions around whales.'"

At public meetings to discuss possible regulations, however, some whale-watching operators were hostile. "The noise is not a new thing for the whales; the whales have been around noise," one of them told NOAA officials. "I just have a hard time believing that these whales cannot conduct their daily lives with the activity."

While the questioner was eager to dismiss the science behind the issue, his remarks reflected a real division within the whale-watching community over the question: How do we watch whales? At one end of the spectrum are activists like Orca Relief Citizens Alliance, whose leaders believe the whale-watching flotilla is actually killing orcas and argues vehemently for a ban on the activity in San Juan County; the Alliance simultaneously advocated land-based whale watching as a low-impact alternative. Perhaps at the other end of the spectrum is Ken Balcomb, the respected orca researcher on San Juan Island, who has been instrumental in cataloging the rise and fall of the resident orcas for some thirty years. Balcomb is skeptical about the effects of the boats on whales' survival, especially since he fears the ensuing fight over it might obscure the real core of the orca problem, namely, a lack of salmon.

He is not a fan of David Bain's research: "I just think he's nuts. It just doesn't make sense. It makes no sense. First, he's making an assumption that that's a significant stressor. I've seen them in vessel stress situations way in excess of what David Bain has seen, and that was at a time when there were fish. They weren't having the population problems they are now. We're going to see some more population drop. For the next twenty years, everybody's gonna wring their hands, but one thing that's for sure is that the human population will increase in that twenty years, and the shipping traffic, the resource exploitation, the building of fuel docks—the system's going to keep on going, and the pressure's going to increase."

As far as Balcomb is concerned, all of the salmon-recovery plans on the table fall woefully short of the orcas' needs, especially if sport and commercial fishing in the Sound continue at their current pace. "There's a lot of talk now about Puget Sound Chinook recovery—there's a plan and the goal is for a few hundred thousand fish. From a fisheries manager's standpoint, if they can get a couple hundred thousand fish, they'd be happy as clams. But that won't sustain these whales. They can go through that in a summer."

Moreover, any salmon-recovery plan won't really take effect until well into the future, at which point it may be too late for the orcas. Many salmon advocates, for instance, are eagerly awaiting the return of healthy runs on the Elwha River on the Olympic Peninsula as a potential fresh source of plentiful orca food following the removal of two salmon-killing dams on the river. But even then, they don't foresee the Elwha producing enough salmon to affect the whales' diet for at least 20 years.

"The recovery goal is way in the future," says Balcomb. "If those fish were here tomorrow, it'd be great, but that's not enough."

For his part, Bain notes that, in reality, he and Balcomb are not so far apart; his work actually demonstrates that any vessel effects are absolutely dependent on salmon abundance. "There's some modeling work that I've done that tries to convert these energy effects that we've measured into population effects, and what that modeling work shows is that when fish abundance is high, there should be no population effects. But when fish abundance is low, the population effects should be strong. So in my mind it's a combination of the low fish abundance, combined with high levels of vessel traffic, that caused the decline. And then vessel traffic levels off, and the fish abundance went up, and that allowed the killer whale population to start to recover."

He also notes that commercial whale-watch operators, although nervous about his research at first, have become comfortable with his findings. "You know, I'm willing to say that this effect is small, and it's something we can do something about quickly, but it's not going to result in any huge population increase. And if we want to get killer whales back to numbers that don't merit an endangered-species listing, the big effort needs to go into restoring salmon. If the salmon abundance is high, then the vessel activity doesn't matter on the life-and-death standpoint. It's a synergistic effect. The two together are more of a problem than either one is alone."

Research released in the summer of 2014 seemed to support Balcomb's contention that the whale-watching boats are not stressing the whales to a serious extent. A University of Washington study, using scat samples collected by a boat that trailed behind the orcas and sniffed out the floating excrement with the help of a sharp-nosed dog named Tucker, found that the whales' stress levels, indicated by levels of glucocorticoids and thyroid hormones, are highest when the whales first arrive in spring, which happens to be a time when whale-watch boats are abundant. Likewise, their stress levels are lowest in late August, when the boats are also numerous. Those stress levels rose and fell not with the presence of boats, but with abundance—or lack thereof—of salmon.

• • •

Most whale-watch captains, including Ivan Reiff, the bearded, forty-something owner of the *Western Prince*, a genial man with twinkling eyes, can see the regulations coming and, for the most part, seem resigned to them. For veteran captains like Reiff, people who believe they're doing the right thing and want to help, not harm, the whales, it's a no-brainer. If the whales go extinct, they won't have much left in the way of business.

"I think it's clear that regulations are on the way," Reiff says as the *Western Prince*, a solid, white-and-red, 46-foot craft with comfortable seating for medium-sized tour groups, chugs around Johns Island.

Captain of the Western Prince, *Ivan Reiff, at the helm.*

"We've already started the process. In a lot of ways, it's just kind of the obvious next step to go, with them being listed. But you hope that the regulations become a tool to really deal with the people who have a blatant disregard for the whales. The worst thing that I've seen is some of the private boats just running right through the middle of the pods. It just makes you cringe. On the one hand, I feel like the whales have enough sense of what's going on that the chances of a boat hitting them is very slim. Part of the reason I say that is I've seen so many times when private boats run right through the middle of them that I think, 'Hey, if they didn't hit them, then they're probably not going to, because that was about as good of a chance of hitting a whale as you'll ever get.' So you hope that the whales know how to deal with it, but it's still just not a good thing. It isn't right. And that is really disheartening, seeing people just totally disregard them.

"Obviously, we have to do what's in the whales' best interests. The whole thing would be a lot easier if it was clear cut, as far as boats and whales in this area. You've got a lot of information out there, a lot of studies, and they're still arguing."

David Bain thinks licensing the whale-watching boats would be a significant step in resolving the issues, since it would give regulators real leverage in stopping bad behavior. "It would also give you a way to cap the number of vessels and give you a way to cap the cumulative effects," Bain notes. "It would in essence put a cap on noise levels, because you won't have noise from different boats being added together. It would also put a cap on exhaust emissions. We don't know if air quality is a factor or not, but it is something we should be thinking about."

The end result may be fewer boats being allowed out in the orcas' waters, as well as rules regarding how many times they may visit the whales and how loud or polluting their engines can be. That likely will affect everyone on the water: commercial operators, private boaters, even kayakers.

The only whale-watching option that won't be affected will be the land-based kind. San Juan Island, in fact, has some of the best options available for that, especially in the vicinity of Lime Kiln Lighthouse on the island's west side, where the state park and the county's Land Bank holdings have reserved a long swath of coastline for watching whales. As it happens, it is also one of the best places to see whales, period. The orcas are known to come in right next to the shore along the whole stretch of land and give visitors an up-close look that they might not get out on a boat.

Kari Koski thinks that the whole spectrum of whale watching has a place in letting people experience seeing killer whales in the wild, but it isn't easy for consumers to figure out the right thing to do.

"I think when people come into the Whale Museum and say, 'Hey, we really want to see whales, we're here from wherever, and it's our dream and we want to do it, how should we do it?' You know, you really need to sit down with those people and figure out what's the best option for them. I think if you look at the spectrum, there's no argument that shore-based whale watching is going to be the least impact to the whales—and some of the best viewing, too.

"We've got fabulous opportunities that not many places in the world do. We've got national parks, state parks, county parks, and our land, all open to the public in premier whale-watching areas that are unbeatable. It's cheap, it's free, you could go down to the county park down to the national park at South Beach and move your way all the way up to Lime Kiln and the county park and watch the whales as they move through and have a fabulous experience."

The key with shore-based watching is that it can be time-consuming and requires patience; you may find yourself parked on a picnic table for a couple of days before whales wander by. (It's what the local residents call "island time.")

"Of the whole spectrum, land-based whale watching is fabulous," says Koski. "You're interacting on your own terms as a human in your own element, when you stand there on the rocks looking at the kelp and there they go; it's a pretty thrilling experience."

Still, some visitors may have limits on their time or children in tow for whom sitting on a grass bank for hours is not an option. "Then the next step is if you want to see them, then we probably would recommend that people go out with a commercial operator," says Koski.

But that step requires some discretion, at least if you're concerned about the whales' well-being. "It's important to ask the right questions when you make your reservations, saying, 'Hey, it's really important to us that you're part of this association. And it's really important to us that you're following the guidelines. You say on your company label that you do, and that's why we want to go with you.' And then if you don't feel like they're doing it, you speak up."

The whale-watching boats, for the most part, are captained by people like Ivan Reiff, who take their role in educating their customers seriously. Reiff sees it as a kind of outreach not just for the whales, but for the whole Puget Sound ecosystem. His trips always emphasize the whole spectrum of wildlife in the islands, from seabirds to seals to minks, otters and eagles.

"I wouldn't feel right about what I'm doing unless it had some purpose," Reiff says, steering the boat around Spieden Island and back towards Friday Harbor. "I've thought long and hard about it, and I've thought, well OK, you can invent a purpose and feel good about what you're doing, and I don't want to do that either. I don't want to pretend like we're doing a good thing and not actually be doing a good thing.

"I do think that we educate people, and that we endear people to these animals and that we are in some way helping to enable their survival. People may not buy solely organic and stop using fertilizer and all that stuff, but at least it comes into their mind, especially when they go to vote, when it comes time to choose people who really have the control over what we do with our environment, you hope that's when it comes into play."

At times, the customers are known to revolt, especially if they've conditioned themselves to expect whales. John Boyd, one of Reiff's naturalists, recalls how, at a previous employer, some Girl Scout troop leaders from Montana, upon discovering that whales were not on the day's likely agenda, went on an obscenity-laced tirade that culminated in their onboard removal by the captain while the boat was still in port.

However, for the most part, even though there's always some disappointment at not seeing whales, customers aboard the boat tours are satisfied with just the chance to take in the abundance of wildlife and scenery. Deborah Smith of Towson, Maryland, was still soaking up the scenery as the *Western Prince* chugged back into Friday Harbor on one of the days when no whales were to be found. The friends from Mercer Island with whom she was staying had told her she'd enjoy the trip. She and a friend took a floatplane from Seattle to San Juan Island and then had been treated to an otherwise full menu of San Juan wildlife.

"I have not seen one thing that I did not like," she said. "I've loved everything. I'm just in awe. We did want to see the whales, but we loved it, and we'll be back."

• • •

In calm seas off the west side of San Juan Island, my kayak bobs gently in a kelp bed. In the water about a quarter-mile distant orcas mill and frolic, most likely hunting their favorite Chinook salmon. I drop my hydrophone in the water to listen to their distinct calls. A loud, low clanging—*whang, whang, whang*—fills my headphones. It is the steady and overpowering sound of a cargo ship, one of the regular features of underwater life in the San Juan Islands' Haro Strait. At first, a quick scan of the horizon doesn't reveal the source of the noise. There are no whale-watching boats around and no large ships immediately visible. Finally, I spot it: a lone log-bearing ship, just visible in the distant haze, heading out to the open sea around the very southern tip of Vancouver Island. *Whang, whang, whang.* It is at least nine miles away. Finally, the ship rounds the bend, and the sea quiets for just a moment before the orcas' distinct whistles, grunts, and rat-a-tat-tat-tats fill the water. These are J pod whales from the Salish Sea's famous and endangered Southern Resident orcas, and they are making a lot of the well-known calls known as "S1."

Seemingly energized, the whales head toward my kelp bed and surround it, chatting loudly and rolling in the kelp. It crackles and pops underwater as the orcas rip up fronds while "kelping" themselves, something the Southern Residents are fond of doing, apparently for the massaging effect. It is one of those moments of pure delight, the sort of take-your-breath-away experience, that comes with being in the territory of wild orcas. At the same time, the ship and its noise cast a shadow, a reminder that the recovery of these endangered whales is precarious, that lingers even as the orcas swim away from the kelp bed and, gradually, out of sight.

The close proximity of freight ships, including those bearing oil, poses a constant risk to killer whales.

The orcas, we know, face many challenges, especially the availability of salmon, to their survival in the interior waters of the Salish Sea, but their ability to hunt for their primary food source is also affected by the noise thrown up by the engines and propellers of various boats. Much of the official focus has been on smaller boats, especially the whale-watching flotilla, and there has been very little attention paid to the effects of these enormous ships. While some researchers such as David Bain may dismiss the effects of big-ship noise on orcas' hunting and communications capacities, because their sounds are at a lower bandwidth than those used by killer whales, other researchers are not so certain. What's particularly noteworthy about large-vessel noise is how overpoweringly loud it can be as well as its long duration; while whale-watching boats may come and go in brief intervals, loud ship noise in Haro Strait can last relentlessly for as long as an hour.

Some researchers fear that, even more than with small boats, the underwater cacophony created by large ships may affect not only killer whales' ability to communicate with each other but more fundamentally may interfere with their ability to hunt successfully. These concerns gain greater urgency when you consider what is on the drawing board for the port in Vancouver, Canada, the source of all the shipping activity in Haro Strait: more ships, many more ships, and among them tankers carrying a black sticky goo known as tar-sands oil.

• • •

Even when ships are present, the view that most people observe across Haro Strait, along the western shore of San Juan Island, when the killer whales are present, is generally a placid one. The only noise is the sounds of the currents rushing, the "*kooosh*" of the whales as they surface and blow plumes into the air, although at times the calm is broken by the engines of the boats that come crowding around the whales to get a close look at them, sometimes thirty at a time. If there are large ships in view, they are mostly distant and seem almost silent as they glide past.

However, drop a hydrophone into those same waters to listen to the sounds in which those whales swim, socialize, and hunt, and the picture changes dramatically. There will be the whales, all talking in their distinct Southern Resident dialect to each other and echolocating for fish. There will be the whale-watching boats, whose engines are mostly short-lived whines and low-level thrums. And then there will be the large cargo ships. Because sound travels so well in water, even the quietest of these will throw up 100 decibels of noise across miles of seawater, and that noise dominates everything within the soundscape. The loudest of them, often older ships with rusty or warped screws, will make noise as loud as 130 decibels, worse than a chainsaw or a turbo-fan aircraft at takeoff.

Val Veirs, a retired physics professor, listens to all this racket, and monitors and records it, with an array of hydrophones he has set up off his waterfront home on the western side of San Juan Island. What he and his fellow scientists (including his son Scott) have found is that killer whales often respond to increased ship noise by vocalizing more loudly, depending on how loud the passing ships are. It is also not uncommon for whales to fall completely silent when passing ships throw up that overwhelming thrum in the water.

According to Veirs, it is inevitable that these noise levels are going to affect orcas, particularly their abilities to communicate and hunt, both of which are closely connected to sound. The noise is well known as an issue for orcas and has been studied many years by different scientists. Veirs, unlike Dave Bain, does not think the shipping noise is inconsequential simply because it is at a lower frequency than the orcas' communications bandwidths. Rather the opposite: Veirs believes the big ships are a significantly greater problem for the whales, primarily because the noise is so loud, so pervasive, and so persistent.

"A whale that makes a 165-decibel call can probably be heard by another whale that's up to a kilometer or so away, under some average conditions," Veirs says. "Under quiet conditions, it could be much farther. And under noisy conditions, much less. So if we say that the average background is about 100 decibels, then around 1 or 2 kilometers is the kind of maximum range you could imagine orcas could communicate. They might do better by increasing their volume, saying, 'Dammit, I'm talking to you over here!' But that gives you a handle on what the range might be for orca vocalization.

"Echolocation is much, much less because there's a huge loss in the echo. There's a 20- or 30-dB loss in bouncing your sound off the fish. So the range at which you can detect your echo is also significantly affected by the amount of underwater noise. So if you go to 125 decibels, which is loud for a ship—but there are 5 percent that go by that are at least that loud—then the distance that this 165-decibel sound could be heard is brought down to about 100 meters. And of course if the whale swims closer to the ship, the sounds are louder, and the distance that they can communicate is presumably even less."

Veirs says the ship noise potential for all this traffic worries him when it comes to the whales. "Right now, about 60 percent of the time, there's no ship within hearing range," he says. "But if you put a couple thousand more ships per year in there, it seems to me you'll end up with about 30 percent of the time instead of 60 percent of the time that the whales are able to communicate without interference from vessel noise."

And even more hair-raising is the knowledge that, if all goes according to the plan of the oil industry, increasingly among those ships will be large tankers bearing tar-sands oil.

• • •

Beginning in early 2014, energy companies started lobbying the Canadian government for the right to ship large quantities of tar-sands oil down the same path these whales take in the summertime. If they win approval, the oil will be carried in an endless parade of noisy oil tankers, negotiating the frequently treacherous currents and hard-right angles of the San Juan/Gulf Islands archipelago alongside threatened orca pods, and what the noise begins, a spill could finish, dooming these orcas to extinction in a few short years.

"Because the whales tend to group up, a catastrophic event such as a spill or a disease outbreak or something like that has the single largest potential to extirpate the population," says Fred Felleman of Seattle, a marine traffic consultant and longtime whale researcher and activist. "We're looking at putting at risk every primary resource that keeps resident whales resident. One significant spill in any one of those areas is more than enough to break the back on a very, very delicate camel."

The placid Salish Sea waters in which the orcas' battle for survival is being played out may seem a long way from Alberta where tar sands are mined, but they are inextricably linked by a pipeline carrying something called dilbit. It begins as a tarry, sludge-like form of petroleum known as bitumen that is extracted from sandstone ore in the Athabasca oil-sands region. Gasoline and other oil products are made from bitumen at refineries. However, there are no refineries in Canada capable of handling it. So the bitumen is diluted (hence "dilbit") with a concoction of highly volatile chemicals into a more fluid form that can be transported by pipeline and then by ship to refineries elsewhere. At least, that's the plan in Vancouver.

Unlike Keystone XL, the more notorious project to transport dilbit from Alberta by pipeline straight to refineries in Texas and the center of an ongoing environmental controversy in Washington, D.C., the Vancouver plan does not cross American soil at all. Rather, the owners of the already-existing Trans Mountain pipeline, that runs from Edmonton, Alberta, to Burnaby, just east of Vancouver, want to expand and upgrade it, increasing its capacity so that large quantities of dilbit can be transported from Alberta to Vancouver, then pumped into waiting ships and transported to various refineries in the United States and elsewhere.

It may surprise most people in the Northwest to learn that Vancouver has become an oil port, not to mention that Haro Strait has become the alternative route to the tar-sands pipeline so many are fighting in the Midwest. The change occurred in 2006 with very little fanfare, when Kinder Morgan, the Houston-based company that owns the Trans Mountain pipeline, began regularly loading dilbit into ships (about five per month) at Westridge Marine Terminal in Vancouver. Up until then, the pipeline had almost solely carried petroleum products headed for use in Vancouver.

Kinder Morgan's long-term expansion plans include tripling the capacity of the pipeline, which would lead to a massive increase in both the amount of dilbit flowing out of the terminal at Westridge and the number of ships carrying it. If the company's plans are approved by the

Canadian Cabinet (and the likely legal challenges fail), Vancouver will see a leap from 60 crude-carrying ships a year to 420 of them. All of these ships can only pass one way en route to the open sea: through Haro Strait, along the American border, and right through the summer hunting grounds of the Southern Resident orcas. Nor is that the entire picture. Overall, the increase in oil-bearing ships represents only about 15 percent of the proposed overall increase in shipping, which includes more container and coal-bearing ships through Haro Strait.

Then there is the controversial proposal to turn Cherry Point, near Bellingham, into a major port for exporting coal brought there by train. The ships coming in and out of that port do not travel through Haro Strait, but rather through Rosario Strait, a narrower and even more convoluted path that also happens to be a regular part of the orcas' circuit-like route.

All of which means, other risks aside, that the waters of the Salish Sea are going to be very, very noisy. However, while a massive increase in ship noise poses a kind of existential threat to the Southern Residents insofar as it may interfere with their ability to find and eat salmon, it pales in comparison to the lethal potential of a possible oil spill in Haro Strait while the whales are present. In Prince William Sound, for example, nearly a third of the resident orcas who were exposed to the oil spilled by the *Exxon Valdez* in 1989 wound up dying within the year. One pod of transients was doomed to extinction by the event, but the Alaskan resident whales' numbers were strong enough to eventually rebound from the losses, as they appear to be doing now.

The Southern Residents, however, would not be so fortunate if they lost a third of their population. "The effects would be devastating," says Ken Balcomb. "It would depend, of course, on whether the whales were in the vicinity when the spill occurred."

While the fiercer weather that might precipitate a shipwreck and lead to a spill would likely hit in winter when the whales aren't around as much, the impact would still linger. Balcomb observes: "Even then the toxins would be in the ecosystem for a long, long time, and that would be the coup de grace. And if they were present, it would pretty much doom whatever pods were in the vicinity."

A spill from a tanker carrying tar-sands oil would be especially lethal. Dilbit does not behave like ordinary crude oil. The dilutants, which are highly toxic and extremely explosive, tend to separate very quickly from the bitumen, which means that the air surrounding the spill will be filled with a cloud of highly toxic and flammable gases. Any air-breathing mammals in the vicinity, including oil-response personnel, nearby residents, and of course killer whales, who are exposed to those gases will almost certainly incur serious lung damage if not an agonizing death.

Compounding the fears of a spill is the reality that government officials in the Northwest on both sides of the border are ill prepared for an oil spill if one were to occur, particularly one involving dilbit, to which the current oil-response units simply don't have the equipment to properly respond. (Among its many dangerous characteristics, dilbit is likely to sink to the bottom of the sea immediately, while most oil-boom and response equipment is geared toward floating crude petroleum.)

"It's always been a question, if there were a really serious spill out here, would the federal government be in any sense equipped to deal with the protection of critical resources," says Karen Wristen, executive director of the Living Oceans Foundation in Vancouver, which is organizing to combat the oil-facility expansion. "And the answer to that is no, they never were."

Making the picture especially gloomy is that the likelihood of a spill increases dramatically in the waters of the Salish Sea as the number of ships starts to climb. One study, by a George Washington University professor who specializes in shipping-traffic risk assessment, found that if all of the proposed projects for increasing ship traffic through Haro and Rosario straits come to pass, the likelihood of a collision by the year 2025 increases by 89 percent. The possibility of "oil flow" or a spill increases over 70 percent. In other words, if all these plans are put into action, it likely is a matter of when, not if, there will be an oil spill in these waters.

Wristen fears the broad impact on the planned stream of oil-laden ships, especially given the likelihood of a spill. "It hugely increases the risk to the whales, and we are so completely unprepared to deal with it," she says. "It would devastate the whole ecosystem, and it would take a pretty hard toll on the people at the top of it, as well, the human beings and their economy. There's so much to be lost in the Salish Sea."

Washington State wildlife officials are aware of the possibility of an oil spill in the interim. Don Noviello of the Washington Department of Fish and Wildlife says state officials have prepared plans to deal with the whales in the event of a spill emergency. Those plans include using underwater bells to drive the whales away, as well as other measures such as seal bombs, although he acknowledges that, while they have run drills testing their equipment, none of these measures has been tested in the presence of whales.

Kinder Morgan and Trans Mountain officials submitted their final proposal for the expansion in December 2013, but the project has inspired large protests and remains in hot dispute, so the approval process by the Canadian Cabinet has been dragged down. The political pressure to approve the expansion will be intense, especially as the fight over the Keystone pipeline drags on and the pressure builds for a means to bring the tar-sands oil to market. While Canada's federal government has been very vocal in support of pipeline development in the West, the opposition is also likely to be intense, especially in British Columbia, where First Nations communities and environmentalists are already up in arms, mostly about the risks posed by pipeline spills along the route to Vancouver.

For its part, the U.S. government in the form of the National Marine Fisheries Service, which listed the Southern Residents as endangered in 2005, noting at the time that a catastrophic oil spill could doom them, says it is prepared to respond to such an emergency, but it is remaining mum on the issue of increased vessel noise and higher spill risk from the new Canadian projects as well as the Bellingham coal port. Pipeline and port officials insist that improvements in navigation and spill response will keep wildlife in the shipping lanes safe. Environmentalists are deeply skeptical.

"We've already got a lot of environmental problems that are slowly driving these whales to extinction," says Ken Balcomb. "I look at those 400 or so ships that they plan to run through here, and it's just another 400 of the thousand cuts that the resident whales suffer. Eventually, they add up."

Keiko at his new home at the Oregon Coast Aquarium in 1996.

CHAPTER *Eight*
Freeing Willies

THE SCENE COULD HAVE BEEN THE BIG CLIMAX IN A MOVIE ABOUT alien landings: a surreal setting in a high-tech facility, the black night split by spotlights, all shining on the big creature suspended in the air, a horde of humans watching with bated breath. This, however, was real life. The creature, Keiko, the movie-star killer whale, was arriving at his new $8 million home.

The horde of media people who had descended on the little seaside tourist town of Newport, Oregon, on January 7, 1996, turned on their klieg lights as the darkness deepened, in order to better catch on video for a waiting worldwide audience the saga of the star of *Free Willy*.

The big six-ton orca, his famous dorsal fin flopped over the side of the white sling, glistened black and white as he dangled from a giant crane. The crowd of well-wishers gathered on the private property next to the aquarium chanted his name.

When the crane carried the whale across the edge of the new state-of-the-art pool, the water in his full view for the first time, he chuffed, sending a spout of fine steam into the air. Then he sounded, once, that trademark orca whistle: "Ooooeee!"

Behind the new pool, in the aquarium's other marine mammal habitats, there arose a sudden din: seals, barking loudly, as if protesting their new neighbor's arrival. Some orcas do, after all, eat seals in the wild.

Those were the only dissenting voices to be heard. Outside, the crowd kept chanting and applauding. Once Keiko was suspended a few feet above the pool, his trainers and veterinarians went to work, drawing final blood samples and checking the whale's responses after more than

20 hours in the sling as he was flown by airplane from Mexico City.

The crowd grew restless, and so did the orca. He began flicking his flukes, knocking one trainer off his feet in the process. Finally they gave the signal and dropped him slowly, fully into the water. A trainer hopped on his back to help him move off the sling, and the orca tossed him off and dove into his deep new waters. It was his first step toward freedom, a return to the wild, truly a first for a captive whale.

The crowd cheered. Even the normally stoic media people applauded. Soon Keiko was responding to the trainer's beckon—a hand splashed in the water, just as the boy in *Free Willy* did—and munching happily on the fish they tossed him. It was a scene right out of the movies. An anthemic soundtrack would have fit right in, which would be appropriate, considering that this was a scene born of the movies, created by and paid for by the movies.

The only thing it lacked was a happy ending.

• • •

Free Willy is slightly odd, as family movies go, in a lot of ways. The most obvious is that the child protagonist is not a typical moppet from the suburbs, which is generally the starting point of most Hollywood kids' films. Jesse, played by Jason James Richter, is a troubled street kid who was abandoned by his mother and is having trouble fitting with anyone, including his new foster family. He meets Willy, the titular orca played by Keiko, after vandalizing the whale's tank.

It's also not a very corporate-friendly movie. The villains of the movie are the aquarium's owners, who can't get the unhappy whale to perform and finally decide to let him die in an intentional "accident" so they can collect the insurance money. At this point, of course, Jesse and friends intervene, setting up the climactic and now-iconic scene in which Willy leaps over a seawall to freedom and a reunion with his family. The indelible message of the film is that captivity is bad for orcas and exists only to make money for aquarium owners.

No doubt this villainous portrayal of marine-park owners, as well as the storyline depicting the freeing of a captive orca, had a lot to do with why, when the film's producers first approached officials at parks such as SeaWorld and the Miami Seaquarium, they were turned away. They finally succeeded with a park in Mexico City called Reino Aventura, which owned a 16-year-old orca taken from Icelandic waters in 1979 when he was three.

Originally dubbed "Sigi" at his first human home in a Reykjavik aquarium, he was renamed "Keiko," a Japanese name meaning "lucky one," when he was sold in 1982 to Marineland in Ontario. Three years later, having developed lesions and showing signs of poor health, he was shipped off to Mexico City. Reino Aventura was originally designed as a dolphin facility, and so the tank in which Keiko was kept was unnaturally small and shallow; when he floated at the surface, his flukes would brush the pool's bottom. His only company was the dolphins owned by the park. He languished there, his lesions and his health worsening; his strength was so poor, he could only just half-breach out of the water. He chewed on the side of the tank and wore down his teeth. It seemed clear it would only be a matter of time for the whale.

So when the Hollywood filmmakers approached the park's owners about their project, any scruples about the script probably vanished with the opportunity to at least make a large chunk of money from Keiko before he died and, at best, to possibly get him into a better situation and forestall that outcome. None of them were quite prepared for the film's overwhelming success.

What really set *Free Willy* apart from the stock boy-and-his-critter story was the animal: Keiko was a surprisingly charismatic animal actor, chosen partly because of his reputation for being easy to work with. Keiko was more than a Flicka or Lassie; the film's director captured accurately the orca's affinity for humans, the affectionate and clever sides of his personality. When, in the course of the plot, Willy saves the boy's life by lifting him unconscious from the bottom of the pool, the scene has an element of credibility lacking from similar scenes in other films. It's easy to believe an orca would be intelligent, caring, and capable enough to make the rescue. Nor was this by accident: A few years before, Keiko had in real life rescued an 18-month-old boy, the son of Reino Aventura's groundskeeper, in nearly the same fashion shown in the movie. The screenwriters heard about the incident and wrote it into the film; Keiko easily recreated the rescue on the first take, which is what you see in the movie.

Audiences may have been suckered by the movie's sappily heart-warming storyline (yes, Jesse does finally bond with his foster family), but the real attraction was the charismatic orca; like the boy in the movie, viewers, especially younger ones, felt connected to the big, smiling creature. That bond manifested itself when, later that year, magazine and newspaper stories examined Keiko's real-life situation at Reino Aventura. A public outcry arose when a number of articles and TV news reports made it known that Keiko's tiny pool could not even filter out the wastes the whale produced daily; that he suffered from skin lesions worsened by the little pool and the polluted Mexico City air; and that fear of contaminating other orcas with the virus causing the lesions was keeping Keiko from being placed in a healthier situation.

At first, it appeared that a team of scientists, led by Ken Balcomb of the Center for Whale Research, would successfully organize a plan to rehabilitate Keiko in an open sea pen in his Icelandic home waters. His owners at Reino Aventura even signed an agreement with Balcomb to do so. Behaving like the villainous owner in the movie, however, the big money behind the captive-orca industry quickly intervened to keep Keiko where he was. Far from eager to see any captive orca ever go free, the Alliance of Marine Mammal Parks, a coalition of marine-park owners dominated by SeaWorld, applied pressure on Reino's owners. Two days after the agreement with Balcomb was announced, it was scuttled when the Alliance announced it would oversee Keiko's care and return him to the wild. A few months later, the Alliance admitted it had no such plan in place, but their intervention was enough to end Balcomb's involvement, as intended.

Keiko remained in limbo until early 1994, when the Earth Island Institute, famous for its campaigns to stop fishing practices that killed dolphins, was named to lead the campaign to help the whale. The Free Willy/Keiko Foundation, headed by Earth Island's Dave Phillips, soon began a series of fund-raisers that would buy a new home for "Willy." Around the world, children began sending in their lunch money and bake-sale dollars for the campaign. One school in Florida alone raised $30,000. In all, some $2 million for Keiko came from children. Their campaigns were often accompanied by children's drawings of leaping, happy orcas.

There were also private discussions going on. Warner Brothers executives realized they would have a public-relations disaster on their hands if "Willy" were to die while in captivity in Mexico City. "Warner Brothers stepped in at this point, and said, 'You know, we need to do something for this whale—we found him, we made him a star, we can't just leave him there,'" recalled Naomi Rose of the Humane Society of the United States (HSUS), who had been involved in the original search and negotiations for the making of the film. This sparked the negotiations that brought about the agreement to move Keiko to a new facility that would be built in Newport. The remaining $6 million needed to make it happen came from donors like Warner Brothers, the Humane Society, and Craig and Wendy McCaw of McCaw Cellular.

The campaign culminated with Keiko's dramatic arrival in Newport. United Parcel Service donated its planes and crews to fly the big whale up by troop transport from Mexico City to Oregon. The Discovery Channel paid for exclusive rights to film the flight for a documentary about the orca. Viewers from around the world tuned in to watch that night's media byte: "Willy" arriving at his new home.

· · ·

Oregon's winter weather is doing its best to train Keiko for a return to the wild. In the first month after his arrival, a fierce hailstorm hit town, then a cold snap iced up the coast for the first time in years, followed by a week of stinging, drenching rain. A pretty good imitation of Iceland, but Keiko is not exactly drinking it all in. He's grown used to lounging in the Mexico City sun. His trainer of the last five years, Karla Corral, has watched him these weeks dealing with the weather: "He doesn't like the wind or the rain very much, I think," she says. "He's ducking under a lot more." Some of that behavior is Keiko's newfound preference for staying underwater. It has now been two months since he arrived in Oregon. He explores his new artificial reef, sometimes rubbing his belly on it, and spends a lot of time in front of the viewing windows, looking back at the crowds, thick with children, that have flooded into Newport to get a look at the movie star.

"He hasn't seen people from that vantage point before and was fascinated," says Mike Glenn, the Oregon Coast Aquarium's curator. "It must have seemed like wide-screen TV to him, with all the coming and going and excited conversations."

Of course, staying underwater makes it easy for Keiko to dodge the weather—except when he eats. Karla Corral carries a bucket of fish and squid to the top side of Keiko's pool, an angle the public never sees, and feeds him several times each day. On this day, the heavy rains hit a peak, blowing in on 30-mph winds, drenching anyone who steps outside for longer than a few minutes.

A Mexico native, Corral is not enamored of the Oregon winters, either. But she says she doesn't mind as long as Keiko's happy. She says the whale has been her best friend for the last five years, and she is glad for the chance to ease his transition to his new home. So the feeding

Keiko nuzzles his longtime trainer, Karla Corral.

this morning, in a whistling wind, is not exactly pleasant. At least not until Keiko's smiling head comes popping out to greet Corral.

"Hey, big guy," she coos to him. He opens his mouth expectantly. She tosses in a handful of fish and squid. He chomps down and dives back underwater. She waits for him to return. Soon, they're going through that day's paces. She holds out a rod with a ball on the end; Keiko spyhops up and twirls in the water just below the ball. She holds out an arm; he swims on his side, his flipper waving hello, a familiar response to fans of *Free Willy*. In between, of course, there are frequent gulps of the handfuls of food Karla tosses in. They talk. She bends down and nuzzles him, nose to nose.

"I have to talk to him a lot," says Corral. "I'll sit down at this end and talk to him about things—if I'm feeling sad or nice, what I do in the days, I sit there and talk to him. I think he really understands, because he stays in, he says things to me. He makes these kind of weird faces. He just sits there and listens. That's why he's great."

Keiko has the same effect on his audiences. The big orca's personal warmth is his most distinguishing characteristic. He seems genuinely to like people, and people respond in kind.

Even as Keiko swam for the first time around his new pool, with its expensive ozone-filtered fresh seawater system and its immense space and artificial rubbing reef, pressure began building behind the scenes. If Keiko could go free, scientists and activists wondered, then why not others? They pointed to some prominent cases where orcas were being kept in conditions worse than Keiko's at Reino Aventura, noting that some of these orcas present fewer logistical problems for release. Such questions, however, strike fear deep in the heart of the people who make millions from keeping orcas captive. They fear most of all that the first orca to be returned to the wild would turn the trickle into a flood. They knew that soon after Keiko, or any other whale, successfully went free, their customers would begin asking why Corky and Shamu at SeaWorld shouldn't go likewise.

"Personally, I think the industry is absolutely terrified at the prospect of any captive orca going free," observed John Hall, a marine biologist and orca expert who worked four years at SeaWorld before leaving in disgust.

Public pressure to free the animals likely would cripple, if not destroy, the captive-orca industry. Already, public displeasure has just about shut down the business of capturing orcas in the wild. The Icelandic captures ended in 1989; until very recently, the last recorded mass wild capture had been in 1997 in Japan, when five killer whales were netted in Japanese waters and packed off to Japanese marine parks. (All five of those whales are now dead.) In 2013, though, Russian fishermen captured eight killer whales from the Sea of Okhost; their fate remains unknown. Another wild whale named Morgan was rescued near the Netherlands in 2010 and is now in a marine park in Tenerife, Spain. Public outcry over the captures convinced marine parks to drop the practices, at least overtly. Their attention now is focused on captive breeding, at which orcas have only been somewhat successful. Orcas' birth rate in captivity has been well under 50 percent: 14 of 39 pregnancies have produced calves who survived the first year. (The infant mortality of wild orcas is unknown, but probably about as high.)

The captive breeding program is not without risks: Gudrun, one of SeaWorld's Icelandic females (captured in 1976), died during childbirth in Orlando in 1996. In 2010, Gudrun's captive-born daughter Taima, still at Sea World Orlando, also died during childbirth. Moreover, the gene pool of captive orcas is very limited and, according to one study, is already showing signs of heavy inbreeding. Finally, perhaps the most disturbing aspect of the captive breeding program is the early age at which females are being impregnated, sometimes as young as four or five years. In the wild, females typically do not begin producing offspring until they reach the age of 12, at earliest.

Ending captures meant the world's supply of captive orcas suddenly became very limited and very valuable. The idea of just setting them free in their home waters is laughable to the orcas' owners, who are, after all, businessmen who are perfectly aware that just dolphins and fish will not pack in crowds eager to be dazzled. For that, they need orcas. Even as the captures ended, however, the ethics of keeping orcas in captivity have grown increasingly dubious, particularly as we learn more about their intelligence and their complex social and emotional lives. Then the question becomes: Why not return wild-born orcas to their native waters and pods?

This was the idea that the Keiko Project was bearing out, and it gained legitimacy with each day that he grew healthier and stronger and learned how to hunt fish again.

• • •

The whole idea of the Keiko Project was not just to give Keiko a happy ending, but to prove that it could be done. In the mind's eye, the idea is almost as simple as the end of the movie: the whale, in a Herculean leap, simply jumps over a retaining wall and is free to return to his family. The realities of reintroduction are much more complex.

"Releasing animals back to the wild can work," says marine biologist Greg Bossert of the University of Miami, who has studied both Keiko and Lolita and has long ties to both SeaWorld and the Miami Seaquarium. "It can work quite well, if you take certain criteria and do it within those channels. We've been releasing manatees since the 1970s, and we have a very high release rate when certain criteria are met. Those criteria include length of time under the care of man, whether or not there's disease transmission potential, on and on and on."

It was Bossert's view at the time that neither Keiko nor Lolita were release candidates and that it would be unethical to consider them, and he continues to feel that way about Lolita. Keiko, he said, had three well-known strikes against him: his weak health, embodied in the papilloma virus that caused the skin lesions around his pectoral fins; a lack of hunting skills; and weak language and social skills, both crucial to survival in the wild. Lolita, he said, was too old and had been in captivity too long. At the same time, Bossert couldn't name any captive orcas that he would consider viable candidates, either.

Indeed, even as Keiko was being lowered into that tank in Oregon, industry spokespeople were inveighing against the possibilities of wild release. They pointed to other problems, notably Keiko's unnaturally worn teeth and the reluctance of officials in Iceland to accept him back. They warned that he might have trouble foraging in the depleted fish stocks of the Atlantic.

Some of their fears were transparently melodramatic and oddly un-self-aware. "The public has not been told that the result of this, the release, almost means certain death for this animal," warned Marilee Keef of the Alliance of Marine Mammal Parks (AMMP). "They've made it sound like it's something that can occur easily." She refused to acknowledge, however, that Keiko's continued captivity at Reino Aventura, something the Alliance refused to do anything about, was far more certain to cause his death.

Keiko quickly proved the doubters wrong about his ability to recover his health. Within the first two months, his food intake doubled, his weight rose dramatically, and he shed the skin lesions from the papilloma virus infection that everyone had noticed in *Free Willy*. Much of that was because he became more active in the much larger tank and thus burned more calories. He became much stronger and more muscular and soon was doing full breaches atop his tank, often without even needing his trainers to encourage him. He began taking an interest in, and then eating, live fish released in his tank.

He was a smash hit at the Oregon Coast Aquarium, drawing 1.3 million visitors in 1996, more than doubling its attendance from the previous year. Children and parents alike lined up to ooh and aah at the big whale through the big windows that looked into his tank from below, and Keiko was always visible when there were people present. He would hang in the water above them like a great glowing angel, contemplative and wise. At other times, he would come face to face with the youngest children who lined the windows and would seem to observe them as they observed him.

When the crowds left and he was alone at night, he would watch movies on a television his trainers set up for him at one of the windows. Mostly he liked videos of other orcas. His favorite movie—the only one he would watch from beginning to end—was *Monty Python and the Holy Grail*. He was utterly uninterested in *Free Willy*.

Soon he became so strong and healthy that Keiko Project officials began making plans to take the next step: Moving to a sea pen in Iceland, where he could reacclimate to his cold home waters and learn how to hunt food on his own again. At least, that was the plan. As always with killer whales, profits beckoned temptingly.

Behind the scenes, officials at the Oregon Coast Aquarium had become addicted to the money stream created by Keiko's immense popularity. Quietly, they began working against the orca's release. Someone began padlocking access to the filters for Keiko pool, and soon they became clogged and nonfunctional. Soon enough, he began showing signs of skin infections caused by swimming in his own excrement. The aquarium leaked word that Keiko was sick, blaming the Free Willy/Keiko Foundation that was overseeing the project. When the foundation pointed out that the aquarium had functionally caused the illness, the aquarium in turn claimed that they had only locked up the filters because the foundation's money had been drying up.

When PBS's *Frontline* featured a report on the feud, Craig McCaw was blunt in his assessment: "It is not in the aquarium's interest for Keiko to be free . . . there's a lot of money at stake," he told the reporter. "And we have felt that perhaps the aquarium was not doing everything it could to possibly bring about the release of Keiko to its highest level."

Aquarium director Phyllis Bell replied: "Well, it depends on how—if Keiko's ready to be released or not. We always have supported their goal of releasing Keiko, if it was possible and it that was the best thing for Keiko. . . . So if the best thing is not releasing him, then he's welcome to stay here." Well, *that* was reassuring.

The aquarium's machinations forced the foundation's hand. Reasoning that Keiko was already healthy enough to thrive in an Icelandic sea pen, they began moving up the date for his departure, and as soon as the facility was finished, they made their move. On September 9, 1998, after two and a half years in Oregon, Keiko was loaded onto a C-17 transport and flown to the Icelandic village of Vestmannaeyjar on the island of Heimaey, where his large new sea pen awaited.

Phyllis Bell was so enraged over having been outmaneuvered that she reportedly swore she would never allow another killer whale into her facility and ensured just that by announcing a major construction project for a new facility that would essentially cut up Keiko's gorgeous new tank into sections and create a set of separate displays. The original budget for the project was about $7 million; it ultimately cost about $11 million. Bell took out $2 million in loans to cover the costs without any authorization from her board and was forced to resign in 2002. She eventually served prison time for forgery in the deception. It was also discovered afterward that Bell had diverted $200,000 in gift-shop profits intended for Keiko to pay for her project. The aquarium defaulted on $14 million in revenue bonds and endured a panoply of other hardships caused by Bell's serial malfeasance.

Once in Iceland, Keiko grew even stronger and healthier and seemingly happier. The Free Willy/Keiko Foundation then partnered with Jean-Michael Cousteau's Ocean Futures Society, which would be in charge of overseeing Keiko's care and rehabilitation. However, Ocean Futures then hired a crew mostly comprised of former SeaWorld staffers to run this program. According to Howard Garrett of the Orca Network, the two chief staffers considered "the entire Keiko Project a misbegotten movie-inspired project and an attempt by Craig McCaw to do the impossible," and he warned Naomi Rose at the HSUS that "in effect, SeaWorld has taken over the Keiko Project."

When Rose paid a subsequent visit to Keiko in Iceland, she was impressed by how well the whole operation was being run, and she was especially impressed by Keiko's health, recalling the sickly thing she had first encountered in Mexico City. "It was a completely different whale," she said. Notably, he was not only catching live fish, he was building real stamina, going out on "walks" with his trainers, following a boat out of his pen into the open ocean and then returning with them. On several occasions, they encountered wild killer whales, and Keiko took an interest in them, vocalizing and joining them for brief periods. These were exciting developments.

At this point, the Keiko Project by any fair measure was already a success: Keiko was far stronger and healthier than he ever would have been in tank captivity, including at Oregon. Whatever time he had left alive had already been extended by years, probably many of them, especially compared to what little future he faced back in 1993, when *Free Willy* was released. At the same time, the ultimate goal of reuniting him with his family and returning him to an ordinary life in the wild was in doubt. It was clear to Rose that her handlers were keeping him on an extremely short leash and ultimately were undermining the effort to disconnect Keiko from human contact. She became especially concerned when they told reporters that Keiko's hopes for a return to the wild were "only 50/50 at best."

There certainly was an important missing element in the plan: No one knew anything about any of the Icelandic wild orca societies, since no photo-ID census had ever been attempted. Balcomb says he proposed such a study in the late 1990s and was turned away; since then, there has been more scientific study of the Icelandic orcas, but no one has yet assembled a complete census.

Colin Baird, Keiko's trainer in Iceland and the man who oversaw his release, says there was some data collected in Iceland in an effort to ascertain Keiko's home pod, but the icy and rough-sea conditions around Iceland are such that collecting enough data on North Atlantic orcas to provide a full photo census of the population would have taken years if not decades (as opposed to the Salish Sea orcas, where dozens of scientists have had ample opportunities in a comparatively pleasant setting amid a large human population for the sightings needed to build a useful database). So, according to Baird, Keiko's team decided to see if he could simply unite with any of the resident orca pods, which he eventually did.

However, Craig McCaw soon lost a huge chunk of his massive fortune when the tech bubble burst in 2001, and he had to drop out of the picture altogether. Ocean Futures' money dried up, and the SeaWorld team left the project. Rose's HSUS took over the project, keeping Baird and several local Icelandic staff members aboard.

Keiko had an especially close relationship with Baird, the former Sealand of the Pacific trainer who had handled Tilikum and had helped retrieve Keltie Byrne's body. They frequently swam together, and Baird particularly worked to help Keiko form a bond with wild whales, since he knew that would be key to their ultimate success. The first of those meetings, about a year and a half after Keiko had arrived in Iceland, came about with Baird's help, on a clear and sunny day when the seas were teeming with orcas, a gathering of three pods at one spot. Keiko was just hanging out next to his orange boat, watching them intently, shyly.

"We were sort of waiting and encouraging him to go out as best we could," recalls Baird. "So finally I just put on a suit and jumped in the water with him, said, 'Come on, let's go,' and basically the two of us swam out into the middle of 90-plus orcas. And I think that's when his confidence really shot up. We had a pretty good afternoon."

After a while, he started hanging out with these orcas and chowing down with them; they were all busy feeding on the schools of herring in the area, and Keiko joined in. Eventually, he did not return to his bay pen, but remained in the area with the orcas and would contact the boats sent out to monitor him. Finally, on August 4, 2002, one of the massive, howling storms common to this part of the world interrupted the team's ability to monitor Keiko, and sometime during the storm, the whole wild superpod headed out for parts unknown, swimming northeasterly across the Atlantic. Keiko stayed with them.

A tag attached to his dorsal fin let the team catch up to him a week later, and they found he was still hanging out, albeit peripherally, with the wild orcas, but he eventually broke away from them on his own and headed almost due east. At that point, he swam almost directly to Norway, a 1,000-mile swim that took him another six weeks and which he undertook completely alone. Keiko found his way into a rural fjord called Taknes Bay and in little time made contact with the human residents there in their boats. His team of handlers arrived in Norway shortly afterward, and Keiko promptly recognized them and began staying with them and their boat. They were astonished by how healthy and strong he looked; when Baird measured his girth the next week, it was the same number as when Keiko was at his meatiest in Iceland.

So, it was clear that Keiko not only had reached a point where he could care for himself, but he could thrive doing it. Most of all, his handlers say, he was happy and frisky and seemed content. But orcas are social animals, and since he was unable to hook up with his orca family in the wild, he sought out the human family he had become accustomed to as a replacement, much as Luna had in Nootka Sound. The Norwegians embraced him. Tourists came from all over Europe to little Taknes Bay to see him by the thousands, and as many as fifty boats could be seen on the water at times from people going out to visit him.

Little children were photographed frolicking with him in the shallows.

At that point, Rose said, the team reached the conclusion that Keiko would probably never be successfully reunited with wild whales and would have to remain under a form of human care for the remainder of his years. It was an experimental form of care; instead of the sea pen he had in Iceland, Keiko was free to come and go as he chose. The team rented a little house overlooking the bay and spent their days on the boat with him. They fed him fish as a treat and a form of reward for engaging them, but it was obvious that he was mostly feeding himself.

The experiment worked well for a year and a half. Keiko thrived and was happy, and the sometimes overbearing relationship with Norwegian tourists began to mellow. However, the marine-park industry remained determined to prove the Keiko experiment a failure, and soon spokesmen for the industry began claiming that the release experiment had now officially failed, and it was time to return Keiko to captivity. Indeed, Arthur Hertz of Miami Seaquarium tried in the fall of 2002 to have Keiko captured and placed under his care, claiming he wanted to find a companion for Lolita in her tank at his facility. (He never did explain how he expected two orcas to co-exist in a tiny, cramped, and shallow pool that is even smaller than the one from which Keiko escaped in Mexico City.) Hertz managed to convince the National Marine Fisheries Service to support his efforts.

The Norwegian government cut off this effort at the knees. "Keiko is doing well, and he is getting a lot of support. There is no immediate need for a rescue," they replied, adding: "We are skeptical to keeping huge animals like whales in captivity. … We do not doubt that Keiko would get good support in Miami, but it would be a great step back to put him in an aquarium that will stress him."

Then one day in early December of 2003, his many years in captivity caught up with Keiko. He came down with a lung infection that is common in captive killer whales and almost universally kills them unless it can be diagnosed early. Keiko, however, displayed no symptoms until it was too late. And so on December 12, he quite suddenly passed away. Colin Baird was out of town and got the tragic word in an airport.

There was great mourning in Norway at Taknes Bay. The locals buried him under a large stone cairn there on the shore, and no necropsy was performed. The reaction from industry anti-release partisans was swift and predictable. One critic of the Keiko Project sneered that Keiko had been "probably the most expensive animal in human history" and that the entire effort had been folly. One of the former Ocean Futures trainers, a man named Mark Simmons, accused the HSUS team of "abandoning" Keiko, claiming he had been forced to rely on handouts from the Norwegian fishermen. He later described the project as "perhaps the most compelling case of animal exploitation in history."

The scientists who study whales, however, say Keiko's story vividly demonstrates how *not* to return an orca to the wild. After all, wild orcas are highly social animals and hunters; their ability to remain with their familial pods is the key to their survival. Failing to do the primary research for Keiko—namely, identifying his home pod and giving him a chance to reunite with them—eventually ensured that it would not reach its final goal, although in terms of lengthening Keiko's life and making his last years quality years where he was free to come and go as he chose, the project was a brilliant success.

"My belief is that Keiko would have needed direct contact with members of his immediate family and community in order to fully integrate back into a life in the wild," says Paul Spong. "That did not happen in Iceland, and it is very unlikely that it would have happened in Norway. However, this does not mean that it could not happen, given the appropriate circumstances. Had more been known about Keiko's social background, it would have been far easier to put him in contact with members of his family. I do not believe he met his mother or any siblings or close cousins while he was swimming freely in Icelandic waters. He did meet and interact with other orcas, but they were not his kin, so he did not join them permanently. That said, Keiko did get to experience the feel and sounds of the ocean once again, after being surrounded by barren concrete walls for most of his life, and that, I believe, must have come as a profound relief to him. For me, the simple fact that Keiko died as a free whale spells success for the grand project that brought him home. Deniers will deny, spinners will spin, but they cannot erase or alter this truth."

• • •

Spong knew well whereof he spoke. Only a year before Keiko died, Spong and his OrcaLab colleagues had helped orchestrate an effort to return a (briefly) captive whale to the wild, which had been wildly successful.

Springer was first noticed by Mark Sears, an avid orca watcher who, in his daily life, is the caretaker for the Colman Pool, Seattle's only salt-water public swimming pool, located right on the beach at Lincoln Park in West Seattle. Things are stone quiet there in the winters, when the pool is closed, and that happens to be the time of year that the Southern Residents (J pod particularly) have been known to come down to the southern part of Puget Sound and feed on the salmon runs there. En route, they often pass right by the pool and Sears' residence there. So he is always on the lookout for them.

In January of 2002, a friend who worked for the Washington State Ferry system gave Sears a call. A little killer whale had been showing up for several weeks now at the northern end of Vashon Island, often around the ferry terminal there, and it was alone. This was very unusual. Sears motored out in his little runabout, since it was just a few miles crossing to Vashon from his home. As he pulled up to the bay near the terminal, he dropped his speed and got out his binoculars. Where was this little orca?

Kooosh! Right beside his boat was where. When he had sufficiently recovered from his near heart attack, Sears got out his camera and took some pictures. When he got back home, he gave Ken Balcomb a call and let him know there was now a confirmed sighting of a solo orca calf hanging out near Vashon. The word got out and onto the evening news, and now the little calf was all over it, featured on the front pages of the newspapers and in breathless media reports with footage of the calf frolicking near the ferry dock. It was a little female, about two years old, scientists eventually ascertained. At first, the ferry workers named her

Boo (an acronym for Baby Orphan Orca), but eventually she was given the name Springer.

Like Luna, who was still roaming Nootka Sound at this time, she was a very playful orca who loved to approach humans in their boats to fulfill her social needs. This began causing a lot of concern, however, especially as it appeared she was becoming underweight. She was feeding herself but not well enough. The scientists pored over photographs of her, along with those of wild whales, to see if they could find any identifying marks that would locate her home pod. No one knew if she was a transient or resident or which pod she might have come from.

Then, Joe Olson of Seattle's Cetacean Research Technology went out and dropped in a hydrophone and obtained a quality recording of Springer's calls. The sounds were then relayed around the network of scientists who study the whales to see if anyone could identify her that way. Sure enough: Helena Symonds at OrcaLab, Paul Spong's wife, promptly recognized Springer's calls. They were the distinctive sound made by A45, a female matriarch of the Northern Residents who had gone missing, meaning the calf had to be her offspring, A73. Graeme Ellis ran a quick check to see who had been notably absent that fall from the A pod, and he confirmed that A73 was, in fact, missing.

No one had any idea how she had become separated from her pod (A pod whales never travel south of the strait of Georgia) or how she had found her way to southern Puget Sound, but now they knew who she was and where she belonged. There was one hitch: Her mother, A45, was also missing and presumed dead, a fact that probably played a role in her separation. So, if Springer was going to return to her home pod, it would have to be with relatives and not her natural mother. A45 had been part of a large clan with numerous relatives still alive and well, so this was probably not a serious issue.

However, concern was growing about Springer's behavior. She began displaying Luna-like behavior: approaching boats, rubbing on them (dangerously close to propellers, in some cases), interacting and playing with the humans in them, and showing dependence on humans for her social needs. Monitors tried to keep the crowds away, but there was only so much they could do. In addition, her health seemed to be deteriorating. The scientists knew that this couldn't go on indefinitely and began forming a plan to reunite her with her Northern Resident family.

Soon, they had a plan in place: Springer would be held briefly in a sea pen and returned to health. Then she would be transported to Vancouver Island and cared for until she had a chance to reunite with her home pod. The National Marine Fisheries Service (NMFS) balked at these plans at first, opting not to intervene because it had no funds to undertake such a plan, and besides, it had never been tried before. There had never been a successful re-introduction of a whale back to the wild after the intervention of humans, although there had been several escapes of penned killer whales in the 1960s, whales who successfully returned to the wild on their own.

Finally, a coalition of private groups—notably the Orca Conservancy, a Seattle-based orca-advocacy organization led by Fred Felleman and Michael Harris; the Whale Museum; People for Puget Sound; and Project SeaWolf, whose monitors had spent the most time on the water with Springer—convinced NMFS that they could obtain funding for the plan through a series of grants, and with building public pressure, the government finally announced it would proceed with the proposal. The Vancouver Aquarium was designated the chief Canadian partner for the operation and, after some back and forth, eventually played a critical role in its success. Paul Spong and the OrcaLab crew went to work coordinating the whale's arrival at the northern end.

In June 2002, they were ready. Springer was gently roped and pulled up in a sling and onto a boat, then transported to a little bay not far away, near Manchester, where they held her for four weeks, feeding her live salmon and keeping the human contact to a minimum. On July 13, they loaded her into a catamaran that had been specially adapted to hold the little orca in a sling between the two pontoons. With TV helicopters and boats following, the procession headed out of Puget Sound and up Haro Strait into Canada. At Campbell River, midway up Vancouver Island, the boat pulled over and took on board the hundreds of bags of ice that had been donated by locals as a "welcome home" for the calf, helping to cool her off at a critical juncture in the journey.

Finally, just as evening was approaching, the boat reached its destination: Springer's sea pen on Hanson Island, not far from OrcaLab, which had been stocked the previous day with live salmon caught by members of the local Namgis tribe. It faced out onto Johnstone Strait, and A pod whales had been coming by frequently. Springer began vocalizing loudly. That night, a small pod of wild whales came by, and one of them had a vocal exchange with Springer. When they went away, Springer frolicked and vocalized excitedly.

The next day, the A pod whales, some from the pod led by Springer's grandmother, were there, displaying extreme curiosity about the little orca inside the sea pen, and some of them entered the bay where the pen was and came close. Springer's human team decided to open up the net on her pen to let her out if she so desired. She immediately swam out and toward the wild orcas, who appeared to be waiting for her. There was a brief interaction, and then the A pod whales swam off on their own to the east. Springer went west on her own.

However, the next day, as her monitors followed her in a boat, it became clear that Springer was tagging along behind her grandmother's pod, and over the next several days, she began interacting with them more and more, particularly with an A5 subpod whale, A51, who began acting as a surrogate mother, pushing Springer away from boats when she was tempted to approach them. By late August, it was clear that Springer had fully reestablished herself with the wild whale community.

Paul Spong was jubilant. "There can now be no question about the success of the return project as it is clear that Springer has resumed living a normal social life among her kin and community," he told reporters.

Springer has not only been sighted with these whales every year since, but she has grown up to become a mother herself. In the spring of 2013, Graeme Ellis and John Ford spotted her with her own infant calf, frolicking in the waters near Bella Bella, British Columbia, far north of Vancouver Island.

"I think this clearly is another sign that such rehabilitation is possible," Ford observed.

Graeme Ellis was especially gratified to see Springer's baby: "It was just a relief for me after all these years since her reintroduction," he said. "It's the ultimate success, I think."

• • •

No matter how you felt about captivity, it was always an awe-inspiring sight to see Finna and Bjossa breaching simultaneously out of the water. It made the whole crowd gasp; two giant sea creatures, rising 20 feet together into the air, twisting in unison, and splashing down gracefully a few feet apart.

Finna and Bjossa, breaching for the crowd at Vancouver Aquarium in 1994.

In the years before the Vancouver Aquarium got out of the business of displaying captive orcas, it was a sight you could always pay to go see. If it was a typically chilly and drizzly day at the Vancouver Aquarium, nestled in Stanley Park at the heart of the British Columbia metropolis, most people held umbrellas, trying to stay dry as they watched the display at the aquarium's outdoor orca pool.

Not all of them, however. A large knot of children would stay up close to the pool's edge, in the "splash zone." When the two whales finished their big breach, the giant wave that crashed up over the pool's edge drenched the kids, who would then squeal in delight. It was like being anointed.

In the flesh, orcas are so much more impressive than even the best wildlife film. Their mere physical presence, a combination of mass and grace, surprises even jaundiced adults, and if you make eye contact, those deep, black, wise-looking orbs can feel like they're probing your soul. It's downright unnerving.

"I call killer whales, and big cetaceans in general, the light-switch," says John Nightingale, Vancouver Aquarium's executive director. "I've never seen another animal, maybe short of a giant panda, reaching out and getting ahold of people's core of their brain, for just a moment. What you do with it, then, once you've got ahold of it, is up to the institution, and that's the real struggle."

At Vancouver, the switch often turned on an interest in nature. After watching Finna and Bjossa, kids could tour the aquarium's reef displays, or walk through its simulated rain forest, or watch a beluga-whale mother nursing her newborn. Along the way, they learned about ecosystems and about how human activity threatens these creatures. It's an important, up-close education. Nightingale has always insisted that the aquarium's board be dedicated to education as the best way to promote conservation. Nearly everything at the aquarium backs that up. Every child who visits leaves with a better understanding of the importance of protecting the world's natural balance. The light, more than likely, has been turned on, at least for a while. That alone, Nightingale argues, was reason enough to have captive orcas. "As ambassadors of their kind, they continue to teach new generations to be interested and to care," he says.

However, responding to public pressure, the Aquarium announced in 1996 that it was adopting a policy precluding any further captures of wild orcas. When Finna died of pneumonia in 1997, leaving Bjossa with no companionship besides dolphins, the Aquarium decided to trade Bjossa rather than try to bring in another whale. She was traded to Sea World in San Diego, where she died four months later.

Nightingale continues to defend keeping orcas captive: "Zoos and aquariums provide access and a vital connection to the world of wildlife and our environment, helping to foster an understanding of nature and how it works, and an appreciation for why it matters," he explained in a CNN op-ed titled "The Case for Captive Animals."

At overtly educational facilities like Vancouver, his words might ring true. At places like the Miami Seaquarium, or any SeaWorld facility, it is quite a different story. Taking your family to a SeaWorld facility or to the Seaquarium, is akin going to a nautical circus show; the emphasis is on crowd-pleasing stunts, like the trainers astride Lolita's pectoral fins when she breaches. Trainers routinely ride the backs of the orcas at SeaWorld. Crowds learn only bare essentials about the animals, if they bother to pay any attention amid all the glitz.

"SeaWorld's primary motto always was, 'Entertainment, education, conservation,'" says John Hall, a marine biologist who worked as a researcher for SeaWorld from 1986 to 1990. "And entertainment was always first. If it were a headline, entertainment would have been in 64 point, education might have been in 14, and conservation would have been in 8 point."

If you ask anyone at these facilities about wild killer whales—especially those in the Northwest, where the effects of the captures that founded the industry are still felt by the surviving resident whales—you are given the "scary oceans" line that in the wild, orcas face many threats to their survival, including lack of salmon, pollution, and boats, and they don't face these threats at SeaWorld, so they are better off there. You are also likely to get false information about their longevity. SeaWorld guides like to claim that wild whales only live about 25 or 30 years. (The reality is that they have a lifespan similar to that of humans; the upper limit of age for wild males appears to be about 60, and females have lived to be 100.)

Lolita performs a backwards breach as part of her twice-daily routine at Miami Seaquarium.

SeaWorld officials defend their education outreach programs, saying they've been upgraded in recent years. Industry officials say the company changed its philosophy dramatically after Anheuser-Busch bought it from Harcourt Brace Jovanovich, and it has maintained those standards since being acquired by the Blackstone Group.

Miami Seaquarium officials offer a similar defense of their education work. Still, it's unlikely a child will watch a Lolita show, a glitzy affair with leaping dolphins and Lolita's playful stunts, and come away knowing much relevant about the animal beyond basic physiology. Even the Seaquarium's public-relations director admits the show won't tell kids where Lolita comes from, what life for orcas is like in the wild, what threats face her native L pod. Nothing they will have gleaned from the show would contribute to understanding the conservation issues orcas represent. But then, these are the kinds of facts that prompt children to ask uncomfortable questions like, "Why isn't Lolita with her family, Mommy?"

Even an industry stalwart like John Nightingale is uncomfortable with some of the displays and the emphasis on stunts that are the métier of many captive-orca facilities: "I do have a bone to pick with some institutions. I think that kind of message is counter-productive." However, he defends the stunts and shows, which the Vancouver Aquarium itself disdained, as part of an effort to get people involved.

John Hall, noting that performing routines with unnatural behaviors have been linked to cases of mental instability in the whales, thinks the stunts are a rip-off. "You have to remember that what people see in these shows isn't a killer whale," he contends. "It's a circus act, nothing more."

Nonetheless, for those who suggest that marine parks' sole interest is snagging the $100 or more the average customer drops with each visit, industry officials point to the research funded by the packed stands. SeaWorld, says John Nightingale, is "able to do things, to build things, to provide advances in medical care, that we can't, because of their size and budget. They've provided huge increases in our knowledge base about some aspects of whales and dolphins that you could never get in the wild." Similarly, the Vancouver Aquarium has long funded the ground-breaking work of Northwest marine scientists like John Ford and Graeme Ellis.

However, there may be limits to the science coming out of places like SeaWorld. John Hall recalls watching the company's chief veterinarian blithely dismiss a proposal to do energetics research on wild mammals, as well as other research that didn't directly affect SeaWorld's chief concern of keeping the captives alive and healthy. "Here was this organization that was generating, even then, $5-600 million a year, and they were putting essentially none of it back into really understanding what was going on with these guys, what it took for them to make a living in the wild," says Hall. "Here was this tremendous capability, and they simply weren't using it."

The two sides—scientists inside the industry, and those outside who ally themselves with whale activists—aren't shy about drawing knives to impugn each other's motives, either. Ken Balcomb is frequently portrayed by those in the Alliance camp as a self-important loose cannon: "Money and ego are the motivating factors," says Nightingale. "The activists shout money, money, money at Wometco and the Miami Seaquarium. But Ken Balcomb's whole living is tied up in shaking these trees and raising money to study this and study that."

There's no doubt that Balcomb, whose chief living for many years was made from an Earthwatch grant that subsidized the Center for Whale Research's summer program, runs a well-oiled operation that one researcher called a "propaganda machine." However, scientists who choose to try to make a living outside the establishment usually have to resort to such methods to raise funds for their research.

Industry officials, for their part, like to claim the orcas' best interests comprise their only motives: "The Alliance is a number of public-display facilities that care very much about these animals. That's where we're coming from," says AMMP's Marilee Keef.

Both sides, for that matter, claim to have the animals' welfare foremost in their minds, but the industry's credibility in this regard is thin, to put it charitably. This is, after all, an industry that was founded on cruel animal-capture techniques considered inhumane and illegal today. While orca care in captivity has improved measurably in the recent past, the industry still regularly engages in appalling practices like "whale laundering" or warehousing orcas captured overseas (orcas are not legally available in U.S. waters) at a windowless backroom tank in another nation until sufficient time passes for it to be imported to a U.S. facility as a transfer, all so the American company doesn't have to obtain a U.S. capture permit. The warehouses are also used to house uncooperative or aggressive animals. The warehousing of Junior, an Icelandic orca who was captured in 1986 and languished without daylight in an Ontario warehouse for four years before he finally died in 1994 may be the most atrocious example of this.

Most of all, we now know without question that the fifty-year record of orca captivity has demonstrated that the only upside for captivity is for the humans who are entertained and, perhaps, educated by them and for the humans who make large profits from them. For the whales, it is unremittingly bad news:

- Most wild orcas have died prematurely in captivity. Of the more than 130 orcas who have been captured in the wild since 1964, only 13 remain alive today (not counting the eight orcas recently captured by Russian fishermen, whose fate is currently unknown). Most were captured at relatively young ages, and many likely would still be alive in the wild. In contrast to many animals in zoos and aquariums, who often enjoy comparable if not better lifespans in captivity, killer whales have dramatically lower lifespans in captivity than they do in the wild: Wild whales of all sexes live only an average 8.5 years in captivity, while in the wild the average lifespan is 30 years for males, and 50 for females. Captive-born orcas have shown a high mortality rate as well.

- There is a high rate of lethal infection in killer whales. Many of these are induced by the whales' dental issues, caused by the tendency to chew on metal gates and concrete pools. Even their dental care, when the pulp is simply drilled out of their teeth and the hole left uncapped, causes problems, because food collects in the unfilled holes and creates even more infections and can lead to septic poisoning if allowed to fester. There is also a high rate of respiratory infections.

- There is a high rate of aberrant behavior in captive orcas, particularly among those who are required to perform unnatural behaviors, including the entertaining stunts that wow so many spectators. The captive orcas frequently engage in aggressive and violent behaviors with each other, something rarely if ever observed in the wild. Much of this is believed to be the product of marine parks' tendencies to throw together orcas from disparate social groups and ecotypes (transients thrown in together with residents, Icelandic whales with Northwest orcas) so that the cultural boundaries that have defined these whales' behaviors all their lives is discarded and ignored.

- The only times that humans have ever been injured or killed by orcas have occurred in captivity. In the wild, their relationship with humans has been unremittingly tolerant and genial, but in captivity, people have been mauled, crushed, raked, and killed.

Especially their trainers, as the Tilikum fiasco demonstrated.

• • •

By late 2013, the sensitivities of everyone working in the marine-park industry seemed to be on high alert, thanks to what people were calling "the *Blackfish* effect." There were many people entering their stadiums now asking that touchy question at the core of the national debate the film ignited: Is it ethical to keep large, wild-born killer whales in captivity? For many people who viewed the film, the answer is an obvious and resounding "No." The evidence it presents against captivity is both wide-ranging and scientifically (not to mention morally) persuasive. However, even people who haven't seen the film often come away with the same abiding impression after seeing orcas in pools at SeaWorld in its three locations (Orlando, San Antonio, and San Diego) as well as at the Miami Seaquarium: There is something deeply Not Right about keeping such large, magnificent, and highly intelligent animals in such relatively small and sensorally sterile enclosures.

There were several *Blackfish*-related protests, including an attempt to block a SeaWorld float at the Rose Parade in Pasadena. Fans of the film organized a boycott at the theme park of a rock-concert series, with various well-known rock and country bands, including Willie Nelson, Cheap Trick, Heart, REO Speedwagon, and Martina McBride cancelling performances. Ann Wilson, the lead singer of Heart, told *Rolling Stone*: "What SeaWorld does is slavery."

In the spring of 2013, SeaWorld stock, taken public by Blackstone, the holding company that bought SeaWorld from Anheuser Busch in 2011, plummeted and then lingered in the doldrums well through 2014. At one point, CNBC stock analyst Jim Cramer advised clients to dump their SeaWorld stock after a bad week of missed earnings expectations and advised the company, "Free Willy!"

In the meantime, *Blackfish* garnered increasing attention as one of the most-watched documentaries of the year. It was shortlisted for the Oscars in the documentary competition but did not make the final cut when the five finalists were announced in January. The film's many fans took to Twitter to complain about the snub, and SeaWorld celebrated when its stock spiked higher briefly. However, even as people walked out of theaters after seeing the film determined never to turn the stiles at SeaWorld again, many found themselves asking, What next? Now that they are aware of the problems with killer whales in captivity, what should their next step be? If we want to get serious about returning captive-born whales to the wild, what whale should they focus their efforts on?

The answer, according to most people familiar with orca captivity, is clear: Lolita.

• • •

Lolita is a remarkable whale, and her case is especially striking because of the tiny size of her enclosure, the smallest orca pool in North America. It is a crumbling structure, but its owner cannot improve the situation because of the park's location in the Village of Key Biscayne, where planners are adamantly opposed to any expansion of the park because of their tiny island's innate limitations.

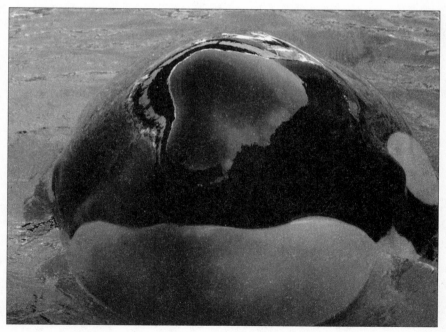

Lolita gladly approaches poolside admirers before her performances.

It is Lolita, more than any other captive orca, who offers the potential to answer the big question that hovered around the *Blackfish* debate: Why not return wild-born orcas to their native waters and pods?

Free Willy and its fantasy of an orca simply leaping over a breakwater to freedom notwithstanding, returning orcas to the wild is not a simple thing. Indeed, as the Keiko saga manifested, it is a complex operation that requires many layers of research and knowledge and physical care of the animals, beginning with identifying any wild orca's natal pod and its location.

Captive whales become inured to human contact and may have real difficulty separating from humans for their social needs as well as for their food. Whales may also carry diseases into the wild, and their dental care (most captive orcas have multiple cavities that require constant maintenance) makes it even more difficult to release them in the wild and expect them to survive. Of the 52 killer whales in captivity at various marine parks around the world (not including eight recently captured in Russian waters), only 13 were born in the wild. Not even the most fervent whale advocates propose releasing captive-born orcas into the wild (although they would argue for improving their living conditions), but keeping wild orcas in concrete pools, they argue, is simply cruel.

Lolita, despite the incredible length of time she has been in captivity (only SeaWorld's Corky, at the San Diego facility, has lived longer in a tank), remains a superb candidate for release. For one thing, her teeth are nearly perfect and her health is terrific. It is hard to find anyone not associated with the marine-park industry who isn't appalled by the tiny size of her enclosure, and many of those associated with the industry will privately admit to deep misgivings about her situation.

Most of all, she is probably atop the list of return candidates because her home pod is well known; indeed, her mother is still living and can be readily located. That is because of the legacy she represents: Lolita is the last surviving whale from the Puget Sound captures of the 1960s and '70s, an episode that provided the foundations for the captive-orca industry. Her capture, in Penn Cove in 1970, is vividly portrayed in *Blackfish*, part of a horrific episode in which five orcas were needlessly killed.

"When I heard the Lolita story, I imagined how amazing it'd be to bring her back to her mother decades after her capture," Gabriela Cowperthwaite, director of *Blackfish*, told me. "This singular, feasible event could catapult us into such a dignified direction. We owe this species big time. And we could start with her."

● ● ●

The average female orca, like Lolita, is about 20 feet long and weighs about 7,000 pounds (males are bigger); the widest part of her pool, at 80 feet, only lets her swim a few body lengths. She needs only two quick flicks of her flukes to travel the span. With her family in the wild, she would swim a hundred miles per day or more, but this is only the most obvious limitation of her captivity. Orcas in the wild are highly social animals, gregarious and playful, whose world revolves around their family pods. They also are complex, large-brained creatures with some sensory capacities, like echolocation, far more sophisticated than humans' own. Holding them permanently in a plain concrete pool is akin to putting a human in a small plain white room.

During the first years of her life, Lolita was a typical Puget Sound Southern Resident orca, feeding on wild salmon, playing with her family, following her mother's lead as her native L pod swam through its home waters. In 1970, when she was probably three or four years old, Lolita was among a large clan of nearly 100 orcas driven into Penn Cove by "orca cowboys." As the young orca's family members lined up in the waters outside the capture and vocalized to the whales inside, the captors selected seven whales to sell to various marine parks around the world and proceeded to lasso them, wrestle them into slings, and lift them out of there. The Miami Seaquarium bought the young whale and named her Lolita.

Since then, her life has been a routine of confinement in a noisy tank that is 30 percent smaller than the tiny Mexico City pool from which millions of school kids "rescued" Keiko. It is the smallest pool for any orca in North America. For the first 10 years, she had the companionship of Hugo, another Southern Resident orca, but since his death in 1980 (he died of an aneurysm after years of bashing his head on the walls of the pool where Lolita still lives, in what has been described by whale activists as a suicide), Lolita has been alone in the tank with only the companionship of dolphins and her human trainers.

Lolita is 47 years old. In the wild, females typically live between 50 and 60 years, sometimes as long as 90. Her presumed mother, the orca known as Ocean Sun, or L-25, is believed to be 85 years old. In captivity, whale activists say, the average lifespan of an orca is eight and a half years (a figure disputed by the industry). Of the 130 or so wild whales who have been captured since 1961, only 13 remain alive today.

Lolita is the last survivor from the dozen years or so that Southern Residents were captured in Puget Sound, a practice that ended with the lawsuit filed by Washington State against SeaWorld in 1975. Those eleven years of captures made possible captive-orca displays like those at SeaWorld and Miami Seaquarium. In the process, some 47 whales were removed from the Southern Resident population, more than a third of the total. The population has never fully recovered; nearly a whole generation of reproduction was represented in the whales captured and killed.

It is that fact that first created a realistic opportunity to pry Lolita out of the hands of her owners. When the National Marine Fisheries Service (NMFS), a division of the National Oceanic and Atmospheric Administration (NOAA), listed the Southern Residents as endangered in 2005, its listing practically singled out Lolita by asserting that orcas from the endangered population already held in captivity—and Lolita is the only surviving such whale—would not be included as part of the population, but following a petition filed by animal advocates, the NMFS decided to reconsider that position. In January 2014, NMFS and NOAA made it official: The agencies announced Lolita would be granted status as a member of the Southern Resident population. What that meant, exactly, was anybody's guess.

For one thing, NOAA has over the years maintained in its rulings that releasing a whale back into the wild could harm both the whale and the endangered population and has been disinclined to support efforts to return Lolita. (Indeed, NMFS sided with Arthur Hertz in his attempts to have Keiko recaptured in Norway and placed back in captivity at his Seaquarium.) The language of the January 2014 finding specified that they would consider simply reintroducing Lolita to the wild a violation of the Endangered Species Act. However, Lynne Barre, at the NMFS, added that the agency was awaiting public comment before deciding exactly the direction they would take in assessing what kind of protections Lolita might be granted under an ESA listing.

"We won't really have an answer to those kinds of questions until we do a full analysis," Barre said, noting that the decision was part of a trend within federal agencies generally to rethink the handling of endangered species of animals.

The immediate impact of the ruling would be relatively limited, since Lolita is the only captive orca from an endangered population in the United States, and it is certainly not clear that a favorable ruling would force the Seaquarium to return her to the Salish Sea. However, it could also potentially affect a number of other animals from endangered populations that are held in American marine parks, zoos, and aquariums as well. Most of all, enabling Lolita's return to her native waters would likely create pressure for other releases as well—at least, if it succeeds. And that's a big if.

Lolita is also at the center of another federal-level case, a lawsuit that centers on a fairly mundane but telling question: How strictly will the federal agency responsible for licensing aquariums, the Animal and Plant Health Inspection Service (APHIS), part of the Department of Agriculture, enforce its own regulatory standards for the industry when dealing with the Miami Seaquarium and its killer-whale pool? A federal lawsuit filed by a coalition of animal-rights groups seeks to force APHIS to enforce regulations regarding the size of enclosures for cetaceans, something APHIS has done with regularity for thirty years, when it comes to licensing Miami Seaquarium. The center of the fight is Lolita's pool or, more specifically, the concrete island that comprises a twenty-foot wall near the center of the pool and provides a platform on which her trainers can stand during shows and onto which Lolita slides and poses at the end of her performances.

APHIS regulations require pools containing orcas to have a "minimum horizontal dimension" (MHD) of at least 48 feet, yet the distance from the island in Lolita's pool to the pool's edge is only 35 feet. APHIS officials, noting that nothing in the regulations explicitly prohibits such a structure, have given the island a "waiver" by asserting that "the platform does not hinder Lolita's ability to move about freely" and is no obstacle to her free movement.

A collection of animal-rights groups and a handful of private citizens filed the lawsuit in the fall of 2011, claiming that APHIS was derelict in its duty to enforce the law, and they now await an initial hearing in federal court in Miami. Emails from APHIS officials obtained as part of the legal proceedings and various Freedom of Information Act (FOIA) requests have revealed that APHIS officials have a deeply contemptuous attitude toward the activists seeking to free Lolita and a strong bias toward the Seaquarium. Indeed, a careful examination of the agency's record makes clear that it is so close to the industry it is supposed to regulate that its rulings, internally, are concerned more with the health of the companies that run aquariums and marine parks than with the health of the animals themselves. Its function, as it has evolved over the years, is more to act as part of the industry's infrastructure rather than to ensure that the industry treats the animals in its care well.

In her nearly twenty years as the marine-mammal specialist for the Humane Society of the United States, Naomi Rose had numerous contacts with APHIS inspectors, directing complaints from their constituents to the federal agency and requesting inspections of suspect facilities. In all that time, she says, only once did the inspection result in any substantive change to the facility.

"Generally speaking, when they have gone in and done the inspection that I've requested, they just give the place a pass, or cite them for something small that they could correct very quickly," Rose says. "Rust on the gate, that sort of thing. And then they get the sign-off. So even though they have been responsive to those requests, they have led to only one substantive action, and that was way back in the 1990s. They certainly never do anything until you write the letter."

• • •

There are a number of other orcas who are prime candidates for release to the wild besides Lolita. Next up on the list would be Morgan, a young female orca held at Loro Parque in the Antilles, and herself the object of considerable controversy (she was rescued, after being separated from her family and found emaciated in the Netherlands, with the proviso that she not be put on public display, a promise her rescuers then ignored). Even though she has been identified as a North Atlantic orca, Morgan's home pod remains uncertain, although John Ford's acoustic research on Morgan provided a close identification of her pod (it's believed she originated with a well-known pod in Norway). This is not the case, however, with another prime candidate for return to the wild, Kshamenk, a mammal-eating orca held in a tiny pool in Argentina—shamefully, it is a tank designed for dolphins that is half the size of Lolita's—who was netted on a beach in an area frequented by his home pod.

More recently, seven killer whales captured last summer by Russian fishermen in the Sea of Okhostk, northeast of Japan, dramatically increased the list of captive orcas. Two of those whales were scheduled for public display at a dolphinarium in Sochi sometime after the closing of the Olympics in February, while the fate of the five other whales is unknown. Whale activists are demanding the release of all seven orcas, since it is believed their natal pods could be readily identified and located.

And then there's Corky, the longest-lived captive, taken in 1969 from the Northern Resident population near northern Vancouver Island. Again, her home pod is well known and easily locatable, and whale scientist Paul Spong of British Columbia's famed OrcaLab, like Ken Balcomb, has a well-developed plan for her return to the wild, beginning with a sea pen, just as he did with Springer. However, as a SeaWorld property, there is little likelihood she will be leaving San Diego anytime soon.

Indeed, the industry remains adamant that none of these whales should be returned to the wild. The Alliance of Marine Mammal Parks asserts (as it has done for years) that "to experts concerned about the risks to which release exposes both the individual animal and the wild population, the issue is a simple one. Without a compelling conservation need such as sustaining a vulnerable species, release may be neither a reasoned approach nor a caring decision."

Nearly everyone involved in orca-rehabilitation issues, however, agrees that the best candidate is Lolita. Ken Balcomb's brother, Howard Garrett, has been running a campaign to "Free Lolita" since the mid-1990s (his Orca Network is part of the APHIS lawsuit). "If anything, we're seeing more energy now than we ever have," Garrett told me recently, acknowledging the role *Blackfish* (in which he appears as an interviewee) has played in that: "When it's placed in front of your eyes like that, people instinctively understand that it's wrong. These animals do not deserve these prisons." Garrett contends that Lolita is the logical whale to be a pioneer because those scientists not only know her family pod's identity and habits, they know that her mother and siblings remain alive. "She's the one," he says. "She deserves to get out of that tank in Miami. She has the strength to do this. She's proven that."

Balcomb and the "Free Lolita" campaign have developed specifics for returning Lolita to her native waters. They want to set up a sea pen at Kanaka Bay off San Juan Island, a place frequented by L pod whales (and not easily accessed by people). From that pen, it is hoped she might be able at least to reestablish contact with her family, if not rejoin them. "We propose to retire her to a sea pen here in the San Juans, where she can at least live out her days in a natural environment," Balcomb told me. "And if she establishes contact with her family, and shows an inclination and ability to hunt and roam free, then we may choose to reunite her with her family. But that's far from a given."

Lolita still uses the signature calls of the L pod orcas in her tank in Miami, and when a *Dateline NBC* reporter played recordings of Southern Resident vocalizations for her, she responded strongly, spyhopping, raising her head above water and peering about and apparently listening to the calls.

None of this concerns Arthur Hertz. Hertz is the president of Wometco, the holding company that, until recently, owned the Seaquarium and the rights to Lolita. Hertz doesn't talk to the press. His PR firm, however, provided me a boilerplate statement about Lolita, insisting: "Moving Lolita in any way, whether to a new pool, a sea pen or to the open waters of the Pacific Northwest, would be an experiment. And it is a risk with her life that we are not willing to take. There is no scientific evidence that the 48-year-old post-reproductive Lolita could survive if she was returned to the ocean." The release further insisted that the facility has always been in compliance with federal rules and that Lolita is "healthy and thriving."

Hertz has refused to even discuss Lolita's sale with the Balcomb group although the offers are in the $2 to $3 million range, according to Howard Garrett. Hertz has even refused to discuss selling her to another aquarium, and there are doubts that SeaWorld would be interested in acquiring her. In the meantime, Lolita's pool continues to age. Damning video footage, taken in the 1990s by Russ Rector of the Dolphin Freedom Foundation and played on local news stations, showed the underside of Lolita's tank at the time: a maze of temporary jacks and supports, rigged to keep the steadily leaking tank bottom in one piece. An algae-covered window looks into the tank. Rector says that when he put his hand on one of these windows as Lolita did her big breach, the glass panel moved a half-inch. The Seaquarium was hit with a number of code violations by city officials in 2003 (mostly involving electrical wiring, some cracked concrete, and a loose railing) and promptly underwent renovations and repairs to address them. No one from the public has seen the pool's underside since then, although the park underwent major upgrades in 2006 after a hurricane damaged parts of the facility, and it had to close.

Hertz promised to expand and improve the pool, but continued to face one hitch: He couldn't. The Seaquarium is located on tiny Key Biscayne, a long spit of sand that connects to Miami by the crowded Rickenbacker Causeway across Biscayne Bay. Key Biscayne residents have clamped down on business expansion and tried to rein in growth and traffic problems. They fear a Seaquarium expansion would worsen their problems. The Village of Key Biscayne successfully fought the Seaquarium's attempts to expand in the 1990s and continues to deny Hertz's efforts to increase parking and improve the pool. In 2003, Hertz told local journalists that he was embarking on a project to expand Lolita's tank, but the project never came to fruition.

A twist in Lolita's story developed in late 2013, when the *Miami Herald* reported that the Seaquarium was in negotiations to sell its facility—and presumably all of its animals—to a California-based theme-park business, Palace Entertainment, that owns facilities across North America as well as in the Antibbes. The sale was completed in March 2014. Whether that means Lolita could end up elsewhere is anyone's guess; the new buyers have declined to comment on any aspect of their acquisition.

A source told the *Miami Herald* that in the recent bid by Palace Entertainment to purchase the Seaquarium, Hertz was offered $30 million for the facility. According to the *Herald*, the Seaquarium's revenues have been rising in recent years despite its aging state. In the meantime, the park's private surveys have shown that two out of three ticket-buyers are coming there to see Lolita. When her daily shows come around, most of the rest of the park goes vacant as everyone packs into her stadium.

Two years ago, there were concerns when Lolita developed dental problems, which can be fatal for captive killer whales. Since then, her health has reportedly improved, but the people working to free her don't know how long all this will last. "We know that whales in captivity only live a few years," says Howard Garrett, head of the Orca Network. "She's already lived far beyond what anyone could expect. For Lolita, it's just a matter of time. They very well could wake up one morning and find her dead. That's what usually happens to these orcas."

• • •

On a typical day, Lolita twice does performances for the crowds who pack the concrete confines of the Miami Seaquarium, once in the early afternoon and again later in the day. These shows are about twenty minutes each and feature acrobatics both from the killer whale and from the Pacific white-sided dolphins (a species nicknamed "lags" by trainers), usually between three and six of them. The other main feature is nonstop blaring rock music of the predictably anthemic variety, rattling around the concrete-and-steel stadium.

Lolita's trainer performing "water work" with her in January 2014.

The dolphins are high leapers and impressively fast, but nothing makes the crowd gasp as Lolita does when she breaches, as she does several times for the crowd. If it's a hot day (as it usually is in Key Biscayne), she splashes the crowds in the first four or five rows with the immense waves she can produce when she breaches (and at one point gives people a complete drenching by splashing them with her powerful tail flukes). The main impression that emerges from all this splashing, however, is that the huge displacements of water created by her leaps and bounds are just like those of a big kid in a bathtub. In this tiny pool, that's about right.

Lolita appears to have an affectionate relationship with her trainers, but, until recently, they have ridden her for years. There was a ride on her side, gripping the right pectoral, around the pool; a ride on her chest, also around the pool; a straight-up breach with the trainer athletically balancing atop her rostrum, high into the air; a similar straight-up-and-down breach with the trainer draped down the whale's chin, gripping the rostrum; and then a standing surfer-style ride around the pool on her back, completed by a brief ride in front of her pectoral onto the pool's slide-out for the show's finale. None of these is a natural behavior for an orca (unlike her breaches, which are commonly seen in Southern Residents like Lolita). Lolita would cap off the show with a giant breach, straight out of the water, with her trainer standing astride her two pectoral fins. However, in the summer of 2014, OSHA, following up on its ruling for SeaWorld, fined the Seaquarium more than $7,000 for permitting the "water work" with Lolita. The Seaquarium says its trainers continue to have physical contact with Lolita, but now that won't occur during performances.

Recent changes notwithstanding, Lolita's routine has remained largely the same for more than forty years. Her original veterinarian, the late Jesse White, described her best: "So courageous and so gentle."

At Miami Seaquarium, they're self-assured that they are doing the right thing for Lolita. Certainly, they deserve credit for having maintained her health well for all these years, although the whale herself also deserves much of the credit for that, too. Nearly everyone who comes in contact with her comes away impressed with her immense internal strength and patience. I was deeply moved by the mutual affection she clearly shared with her trainers, but everyone at the park was reluctant to talk to a reporter from Seattle, perhaps mindful of how their work is depicted in *Blackfish*.

In 2010, Robert Rose, the facility's curator, gave a terse response when an audience member asked him about whether Lolita was happy in that tiny tank as a McClatchy reporter listened: "This is her home.

It's the only home she's known for 40 years." Rose sees a whale's home as a place, but off Kanaka Bay in the summer of 2013, L pod whales, Lolita's pod, including her mother, spent a number of days chatting and milling and munching on Chinook, more or less in the total area of the bay, and they always stayed together. Because for orcas, home isn't just a place. These creatures' home is each other. Wherever their mothers and brothers and aunts and uncles go, usually in search of salmon, that is home. It shifts and drifts with the tide and the fish, but they always have a home.

It's Lolita's real home, too. After forty years and counting, there's still no place like it.

CHAPTER Nine

Orcinus Ethics

T HREE KWAKWAKA'WAKW BOYS WERE OUT IN THEIR CANOES. THEY were of that age when the idea of manhood dangled before them tantalizingly. Not far from the shore near their village, they were out practicing their skills at shooting arrows with a bow and throwing a spear. While they were playing, a pod of killer whales, *Mi'Max'inuxw*, came past, frolicking, splashing. The boys watched in awe for a while, and then one of them started thinking about his manhood.

"I bet I can shoot one of them with my arrow!" he declared to his two friends.

"Me, too!" his friends shouted.

They got out their bows and talked among themselves, deciding to take aim at the orcas' dorsal fins. The other boys missed, but the first boy's arrow found its mark, striking an orca and piercing its dorsal fin. The orca gave a loud *chuff!* and disappeared underwater. Now the whole

A rising orca blows bubbles as it surfaces.

family of killer whales was angry and charged at the three boys. Suddenly stricken with fear, they paddled madly in to shore, arriving one after the other. The boy whose arrow had hit the orca was last, and just as his canoe came to shore, so did a huge killer whale in hot pursuit.

Magically, the whale's dorsal fin transformed into a large man, and he was angry. He reached out and grabbed the boy as he clambered out of his canoe, seizing him by the heel at his Achilles tendon and picking him up. He dangled him that way, pinching the tendon.

"I am *Max'inuxw*! You, boy, as long as you live, will never be able to walk properly, and you will always suffer in pain from the muscles in your heel being pulled out. This is what you did to us. I am *Max'inuxw*."

With that, he let the boy back down, went back to the water and into his canoe, and transformed back to a killer whale. The three boys grew up to lead families that honored the *Max'inuxw*, engraving his image on their doorways and totems and honoring the whales at feasts and potlatches with songs and dances. Most of all, they never allowed anyone to bother or harass the killer whales again. The boy whose arrow struck the whale's fin always walked with a limp after that. Even his descendants paid a price.

"The killer whales are one of the most highly respected creatures in Kwakwaka'wakw culture," says storyteller Andrea Cranmer of Alert Bay, who told this story for a CBC radio audience. "They can take sickness away, guide you to safety, and are regarded as being the same spirit as man. But because of that arrow, that boy's descendants will always have pain in one of their heels, and they will always walk with a limp."

• • •

Can an animal be a person?

That is the essence of the challenge the killer whales pose to humans, especially those humans who hold them captive. But it is also a larger challenge to all of us, especially if we endeavor to take our role as stewards of the world in which we live seriously. It is a strange and alien concept in a world dominated by Western thought, in which humans have historically been regarded as exceptional beings apart from nature and in which all nonhuman occupants of the world are considered animals, at best property and at worst vermin, the extermination of which is required for the sake of human well-beings.

"Right now, there is no one besides a human who is a person," says dolphin scientist and ethicist Lori Marino. "They're all property, no matter how complex they are, no matter how much we love them. They have no inherent rights of their own."

Yet the more we learn about dolphins in general, and killer whales in particular, the more that our assumption of innate superiority looks like a presumption. Orcas, with their big brains, complex social structures, mysterious communications, and mind-boggling sixth sense, by their very existence, challenge the longstanding belief that human beings are the planet's only intelligent occupants. Social life for killer whales, as we have seen, is deeper and more omnipresent than it is for humans; their identities are defined by their families and tribal connections; and their empathy is powerful enough to extend to other species. If orcas have established empathy as a distinctive evolutionary advantage, it might behoove a human race awash in war and psychopathy to pay attention.

We've also learned that these creatures have rich emotional lives. Their brains are extremely developed in the areas associated with emotional learning, and their tight social arrangement, in which family bonds remain for life, is complex and sophisticated. They also have a demonstrated capacity for empathy. Nor, for that matter, is this only true of dolphins and cetaceans generally. The more we learn about a number of creatures that have always been deemed non-persons by dint of their nonhuman status, the more their emotional lives are being revealed: chimpanzees and all the great apes, elephants, even cats and dogs and pigs and cattle, all have more developed emotional centers than we had previously supposed.

Gregory Berns, an Emory University neuroeconomist, has concluded that dogs, for example, provide plenty of food for human thought even beyond what we all thought we knew. "The ability to experience positive emotions, like love and attachment, would mean that dogs have a level of sentience comparable to that of a human child," writes Berns, "and this ability suggests a rethinking of how we treat dogs." Berns, like an increasing number of animal ethicists, contends that this rethinking should take the shape of "a sort of limited personhood for animals that show neurobiological evidence of positive emotions." In other words, the more closely we study animals, the more we find much of what we think of as constituting personhood. It is not exclusive to humans.

This is, of course, sheer craziness to many people. "The whole concept of non-human personhood is fraught, and people respond differently to it," says Lori Marino. "I'm not even sure if personhood is the right word to say, but it's the word we have right now." The high intelligence of certain species—particularly orcas, dolphins, and chimpanzees—has been the driver of the ethicists' own evolution in their views. Most of all, it has forced us to recognize that our definition of intelligence is self-servingly geared to place us on top.

"We ignore the inconvenient fact that we choose to define and measure intelligence in terms of our greatest strengths," observes marine biologist and ethicist Jeff Schweitzer. "We arbitrarily exclude from the definition of intelligence higher brain functions in other animals. Enter the compelling interest in communicating with dolphins. We would be low on the list of smart animals if we included in our basic definition of intelligence the ability to use self-generated sonar to explore the environment and to communicate."

Because personhood has always been associated with intelligence, a less anthropocentric definition of intelligence yields a slightly reconfigured understanding of personhood as well. Marino's dolphin-science colleague, Thomas White, has proposed a definition of personhood that includes being alive, aware, being capable of feeling positive and negative sensation as well as emotions, having a sense of self, having control over one's own behavior, and having the ability to recognize other people and respond appropriately. Most of all,

A person has a variety of sophisticated cognitive abilities. It is capable of analytical, conceptual thought. A person can learn, retain, and recall information. It can solve complex problems with analytical thought. And a person can communicate in a way that suggests thought.

As White explains, dolphins fit this definition more than adequately, as demonstrated in a variety of experiments. Their creativity and inventiveness, for example, were brilliantly exhibited by a female dolphin named Malia at a facility in Hawaii. Malia was rewarded for exhibiting new behaviors and developed an expansive repertoire of stunts beyond anything her trainers had thought possible.

The dolphins' ability to control their behavior and to recognize other individuals is embodied in the interactions with humans observed by dolphin scientist Denise Herzing in southern Florida; those dolphins, in fact, seem to eagerly seek out humans, even though they are not being fed or stroked or otherwise interfered with, and engage them in a variety of play behaviors.

Even more striking is the social dimension of dolphins' and orcas' faculties of perception, especially their echolocation. Just as when that calf I encountered off Kaikash Creek seemed to be listening in on the echolocation bullets from its mother that were striking my kayak, scientists have found that dolphins, too, "eavesdrop" on the echolocation sounds made by their fellow pod members. Brain specialist Harry Jerison observes:

> Intercepted echolocation data could generate objects that are experienced in more nearly the same way by different individuals than ever occurs in communal human experiences when we are passive observers of the same external environment. Since the data are in the auditory domain, the "objects" they generate would be as real as human seen-objects than heard "objects," that are so difficult for us to imagine. They could be vivid natural objects in a dolphin's world.

The "social cognition" that arises from this kind of richly shared experience of the world would even lead to a different sense of self than humans experience. Jerison argues: "The communal experience might actually change the boundaries of the self to include several individuals." This clearly indicates that dolphins—and particularly killer whales, in whom we have observed the most highly developed acoustic skills, as well as the most elaborate social and communicative structures in the delphinid family—have powerful emotional and empathic connections to each other that are integral to their own personal identities as beings in the world. Their togetherness defines them as persons.

White's observations about the personhood of dolphins applies as well, naturally, to orcas, perhaps at an even higher level—meaning that, logically speaking, they qualify for personhood, as do dolphins, and indeed, the same logic thereby opens the door for consideration of the same status for a number of other species of animals.

When we define intelligence in a way that is appropriate to a species, its capabilities, and its environment, that likewise applies to our definitions of personhood. Our traditional definition of personhood is also deeply anthropocentric, based on an experience of the self that encourages highly individualized behaviors. Cetaceans, on the other hand, experience self in a completely different way, one encouraged by an aquatic environment that produces highly social and empathic beings. However, when we start redefining personhood in a less anthropocentric way, there are deep ramifications. That road inevitably leads to the realm of law and legal rights, nominally the province of every person.

"So the rights that proceed from intelligence are species appropriate," says Marino. "You know, dolphin rights are not the same as human rights, and dolphins don't need to have the same rights as humans. So you look at what you can deduce to be what the individual animal needs for a healthy, productive life. And whatever those things are, that is what their needs, their rights are."

The implications proceed beyond just captive cetaceans, however. Many primates are held in captivity in zoos and research facilities around the world. Dogs and cats are considered the property of their owners, to be disposed of as they see fit. So are agricultural animals such as pigs and cows. When we talk about giving them rights, what does that portend for the people who breed and raise the former, and slaughter and eat the latter?

The ethicists say that the rights are naturally limited to what the animals' needs are. Recognizing such animals as "nonhuman persons" doesn't necessarily mean people have to stop eating them or using products from them. It does mean, however, that people would be required to give the animals in their care decent lives in which their daily needs are met. If these ethics were to gain cultural currency, facilities such as the gigantic "pig cities" where animals are raised and die in tiny pens, locked up with thousands of their fellow hogs, would no longer exist.

"I mean, clearly, no one is saying that pigs should have the right to go to college," says Marino. "But they do have the right to be able to move around at will, and be able to have their babies in a way that they like, and those kinds of things. It runs counter to a lot of traditional views, especially the view of animals as property," she acknowledges.

But the science underlying our understanding of them impels a shift in ethics. That shift, she says, is embodied in the growing scientific consensus that captivity is not an appropriate state of being for killer whales. "The bottom line really comes down to the scientific data, which will tell you if these animals can thrive in captivity," Marino says. "And they don't. People ask my opinion all the time, and I say my opinion is irrelevant. Here are the papers that led to my accepting the conclusion that they cannot thrive in captivity. And I always try to impress that upon people.

"They are big and strong and so impressive. But stop and think about how little you really know about them from captivity. What's really impressive about orcas is all the stuff they do in their natural environment—their social life, the way they hunt, the way they travel, the way they partition resources, their cultures—all that stuff, you get no sense of that in captivity. You just get basically the very superficial kind of big giant strong animal splashing in the water."

Marino and a number of her colleagues have joined forces to create the Nonhuman Rights Project (NhRP), an attempt to bring their ethical considerations into the legal realm, actually giving animals rights of their own for the first time. Their first campaign involves giving four captive chimpanzees held at various locales in New York State their relative freedom and moving them to a sanctuary where they could live out their days in a semi-wild environment.

"No one has ever demanded a legal right for a nonhuman animal, until now," said Steven M. Wise, the founder and president of the project, when the lawsuit was announced. "When we go to court on behalf of the first chimpanzee plaintiffs, we'll be asking judges to recognize, for the first time, that these cognitively complex, autonomous beings have the basic legal right to not be imprisoned."

Initially, the lower courts rejected the plaintiffs' arguments, as expected. The attorneys at the project are more hopeful that they can gain traction in the appeals courts. "These were the outcomes we expected," said Wise, after the December 2013 initial ruling in the lawsuit. "All nonhuman animals have been legal things for centuries. That is not going to change easily. . . . The struggle to attain the personhood of such an extraordinarily cognitively complex nonhuman animal as a chimpanzee has barely begun."

Says Marino: "The NhRP is trying to take the first step, which is just to establish common-law legal personhood. It's not even legislative or constitutional. It's just trying to get one judge to say that one chimpanzee is a legal person with a right to bodily liberty. And that's going to be a tall order, but I think eventually we'll get there."

Something similar had been attempted the year before, but instead with orcas. A variety of animal activists, led by People for the Ethical Treatment of Animals, went to court to order SeaWorld to release its killer whales on the grounds that they were being held in "slavery," and thus were in violation of the Constitution. Among the plaintiffs listed were Tilikum, as well as Corky, the last surviving Northern Resident captive orca. However, it was swiftly dismissed by the federal courts, with prejudice.

"The slavery lawsuit didn't succeed for a simple reason, and that was that they wanted the judge to interpret the Constitution in a way that even if he wanted to, he couldn't, which was to see orcas as human slaves," says Lori Marino. "Clearly the 13th Amendment did not provide for orcas. You can go there, but the first thing you have to do is, orcas have to be persons—non-human persons."

The backlash against orca captivity created by *Blackfish*, however, has revived a national discussion of the topic. Increasingly, the public is awakening to the reality that the scientific ethicists have been raising for many years, namely that the more we learn about killer whales, the more we realize that their continued captivity, especially as it is practiced today, is the wrong thing.

• • •

Among the many things we have learned about killer whales in the forty years since we began capturing them and putting them in concrete tanks, probably the most striking is their extremely social natures. The lifelong bonds between mothers and their calves, as well as their fellow pod members, are deep and profound in ways we can only glimpse like fleeting shadows. There is almost no respect for those bonds for orcas in captivity, however, especially when it comes to their breeding program. Young calves are routinely separated from their mothers shortly after birth, in large part because so many of the females in captivity were either captured quite young or were themselves born in captivity and never learned proper nursing skills from their mothers, as they would have in the wild, and so most newborns have to be hand-fed by humans if their owners wish to see them reach adulthood.

Neither is there any respect for the profound cultural differences among orcas. At marine parks, orcas from ecotypes around the globe—. some of them fish-eating resident whales, some of them mammal-eaters—are thrown in together, usually with no consideration for their ecotype of origin, let alone their home pod.

Perhaps most egregious, though, is the stultifying reality of their confinement. Not only are they prevented from swimming the vast distances to which they are accustomed, their pools are invariably feature-less concrete tanks, the likes of which, for creatures used to perceiving the world through sound, is essentially the same as imprisoning humans in small white featureless rooms.

All of these stresses add up to shortened lifespans, unhealthy animals, and most of all unhappy animals, capable of acts of aggression that are unseen among wild killer whales, especially toward their human handlers. Tilikum is far from the only psychologically unstable orca among the ranks of the captives, nor the only one who has acted aggressively against trainers, a reality that *Blackfish* illustrated vividly. The only logical and ethical conclusion is that a new program, creating a new business model, should be undertaken at marine parks: retirement, rehabilitation, and potentially a return to the wild for some whales. It would include captive-born as well as wild-born killer whales, and it would mean an end to all further captive breeding programs.

Naomi Rose, herself a onetime Northwest orca researcher who helped oversee Keiko's final years while with the Humane Society and is now a lead scientist at the Animal Welfare Institute, has proposed a new paradigm for the SeaWorlds of the world: to relinquish their continued use of these animals for entertainment and to instead embrace education and conservation as the centerpieces of their business. "These facilities can work with experts around the world to create sanctuaries where captive orcas can be rehabilitated and retired," explains Rose. "These sanctuaries would be sea pens or netted-off bays or coves, in temperate to cold-water natural habitat. They would offer the animals respite from performing and the constant exposure to a parade of strangers (an entirely unnatural situation for a species whose social groupings are based on family ties and stability; "strangers" essentially do not exist in orca society). Incompatible animals would not be forced to cohabit the same enclosures and family groups would be preserved."

There are many wild-animal sanctuaries in operation around the globe for a variety of species, operating on a nonprofit business model, and Rose argues that such an operation would fit in well even with a business such as SeaWorld, which is driven by profits and ticket sales, so long as it was willing to move to the locales where the rehabilitation was taking place. "Wildlife sanctuaries are sometimes open to the public, although public interaction with the animals is usually minimized," says Rose. "A visitor's center can offer education, real-time remote viewing of the animals, a gift shop, and in the case of whales and dolphins can even be a base for responsible whale watching if the sanctuary is in a suitable location for that activity."

SeaWorld, of course, has so far ignored such importuning as ridiculous and idealistic, while defending its continued practices of holding the animals in captivity. Mostly it does so on dubious grounds. It claims its programs are educational and inspire children, but in reality children are fed a heavy dose of misinformation at these facilities. It claims it gives the animals better and longer lives in captivity, which while true for some species, is decidedly a falsehood when it comes to killer whales; and it claims it contributes to conservation and rescue efforts, but it has done not a single meaningful thing to help restore the Southern Resident population in the Salish Sea on which its industry was founded.

When SeaWorld guides talk about those wild populations at all, it is to offer a stark contrast with the "safe" and sterile environment in which the orcas they possess now live, as though it were obviously preferable to be fed dead herring all day rather than chasing down salmon with one's pod in native waters. They make the oceans out to be scary places rather than the natural and free environments in which these creatures were always meant to live.

There is a powerful element of truth in their observations. Wild killer whales indeed face all kinds of challenges to their sheer survival out there in the ocean. However, that does not mean captivity is a better place for them than the wild. It only means that the people who fear for their well-being in captivity, and wish to see it ended, also need to be engaged in helping killer whales to thrive in the wild.

• • •

Killer whales around the world face a variety of threats. Crozet Island orcas are still being killed by fishermen from whose lines they mischievously steal a whole day's catch. In Russian waters, orcas are being captured and hauled off to inland aquariums, including eight who were captured for display during the 2014 Winter Olympics. Alaskan killer whales are still recovering from the effects of the *Exxon Valdez* oil spill. A shrinking fishery in the North Atlantic continues to stress the populations there.

However, the only known endangered population is the Southern Residents; they have not yet recovered from the effects of losing a generation's worth of young orcas during the capture period of the '60s and '70s. The declining salmon runs, the buildup of toxins and pollutants, the increase in vessels and their accompanying noise have all played a role in increasing their vulnerability and keeping their numbers low. Looming future threats, such as increasing the possibility of a toxic oil spill in their home waters, make their future very cloudy indeed.

Brad Hanson, the NMFS whale scientist, is concerned about the precarious position in which the Southern Residents now find themselves, some of it an outcome of their evolution. "To me, it's fascinating—this adaptation where we see these small, social odontocete populations," Hanson says. "You see it with pseudorca, with killer whales, with belugas, with Baird's beaked whales. When you think about the populations of other animals, which number in the hundreds of thousands or millions or whatever—to me, what it shows is the fragility of these populations. Because there aren't very many of them. And yet they've come to exist in these very small groups and enclaves. The point being, the fact that they do exist in very specialized environments and there are a small number of animals, it's a very fragile situation. And I think the public has a really hard time trying to comprehend all the insults to the environment that we inflict, and in doing so inflict them on these animals and make them even more fragile."

Paul Spong sees the same precarious state: "It's very unfortunate for the Southern Residents that they are in a place that we have chosen to occupy so heavily," he says. However, he fears for the future of both the Southern and Northern Resident whales. "I am concerned for what the future holds for all these orcas, and for all kinds of marine life. I just know that we impact them in all kinds of ways. A lot of it is inadvertent, and a lot of it is quite deliberate.

"Just looking at the issue of pollution, and looking at what it would take to change things around to the point where there would be a significant difference for the orcas, and for other marine mammals. It's massive, because there are so many entry points for toxins into the ocean system."

As Spong explains, these are all issues in which every person who lives in the Northwest can play a role. "You need to start at the personal level," he says. "People in their own ordinary lives can pretty easily make decisions about things that are appropriate and not appropriate."

Hanson agrees, but also sees the complexity: "It's hard for people to understand how to make a difference. They want to help the animals, but they really don't know how," he says. "It's hard to tell people when they say, 'What do we do?' And it's a hard question, because there's so much stuff going on at so many different levels."

For the Southern Residents, those issues run from prey availability to toxins to vessel noise. Do you use yard fertilizers that run off into local waters and harm salmon? Do you go out to see whales in a boat that ignores the self-enforced 100-yard regulatory distance from the orcas? Or do you make conscious choices to do otherwise?

"How does your daily behavior connect to that?" wonders Hanson. "We all have an impact, and it's hard to be conscious of what those impacts are. You go home and you fire up the furnace or you flush the toilet, and it's water from a river system that salmon once depended on. You know, we live in a heavily altered system that sort of marginally supports salmon stocks. But does it make sense to rip out whatever to do stuff to make them strong again? We have to figure out the tradeoffs."

Some of these tradeoffs involve difficult land-use issues. In the Skagit River valley, where a restored river system would probably yield one of the largest bounties of salmon for Puget Sound orcas to devour, in the longtime feuding between farmers, who now occupy much of the river's delta, and tribal and conservationist interests trying to restore salmon runs, a stalemate has ensured that few improvements are in the offing.

That, if anything, reflects the greatest difference between men and orcas, and the trait we would be wisest to learn from them: cooperation. Killer whales are deeply wired to cooperate with one another, to assist each other, to join forces in order to achieve their goals. Our lack of the same trait, ironically, is what in the end most endangers them, and should we ever heed their example and learn that trait, conversely, it could be their salvation.

"There are a lot of ways the killer whales are an indicator species," says San Juan Island naturalist Monika Wieland. "They can tell us a lot about what's going on out there. I mean, the salmon are so crucial for the entire ecosystem here, and the whales are sort of a visual indicator for us of what might be going on out there. You just hope that people are listening to that message—if the whales are not here, especially." She knows that there are reasons, including the whales' disappearing act in the summer of 2013, for pessimism, too. "There's been a lot of doom and gloom," she says. "I know of a lot of naturalists who say things like, 'I don't want to just stay here and watch them die.' People are almost ready to give up on them." She smiles. "I'm not giving up on these guys just yet. Just like the salmon on the Elwha, we know that nature is incredibly resilient. I think if we give them any sort of opportunity, they will take advantage of it."

• • •

Franz Boas, the father of modern anthropology, got his start by spending time among the people who believed that the killer whales were their ancestors, the Kwakwaka'waka. It changed the world.

As a young man, Boas spent a great deal of time, between 1886 and 1890, off and on, in the Northwest, collecting myths and legends and traveling among the various tribes that were scattered along the coastlines, mostly straggling remnants that had managed to survive the onslaught of smallpox and cholera that had nearly destroyed most coastal villages between 1800 and 1870. Over time, he became especially close to the Kwakwaka'waka (whose name, pronounced KWA-kwa-KEW-aka, Boas shortened to Kwakiutl). In 1892, he organized a delegation of fourteen of the tribe's men and women to represent themselves in an exhibit of a mock cedar-longhouse village created for the Chicago World's Fair, where they were gawked at by fairgoers.

Like the killer whales, the Kwakwaka'waka had a matrilineal familial and tribal structures. Boas eventually ascertained that this structure had evolved from patriarchal structures with men at the head, something that ran counter to then-popular theories about how cul-

Franz Boas' photographer captured this image of a Kwakwaka'waka man demonstrating a dance performance in a killer whale transformation mask. Taken in Chicago in 1904 for the Field Museum.

tures "naturally" evolved into patriarchies, which were seen as their highest development. Boas began to challenge these theories, arguing that instead of the biological forces that were popularly believed to drive human behavior, cultural forces were what make us tick.

In the end, after he had returned to Columbia University and founded the study of anthropology as an academic discipline there, the ideas Boas developed during his time among the Kwakiutl not only profoundly shaped academic thought, but they challenged the reigning worldview of the time: white supremacy, and its assorted pseudoscientific manifestations, particularly the fake "science" of racial purity known as eugenics. At the time, it was widely believed that there was a hierarchy of races and civilizations, with Western white society the supreme outcome of evolutionary forces. Tribesmen such as the Kwakwak'awaka were, in this view, hopelessly backward and primitive, scarcely capable of reasoned thought, let alone sophisticated art forms or other cultural expressions. In some eugenicist views, it was not even clear if they were fully human. Boas, himself a Jew who had observed the resemblance of the supremacist worldview to deepening anti-Semitism in his native Europe, had come to know from deep experience that this was utter bosh.

Boas' theories, known popularly today as "multiculturalism," held that cultures cannot be ranked higher or lower, advanced or primitive, superior or inferior; people form these judgments based on the biases inherent from their own cultural learning, he said. At the time, these ideas were widely ridiculed, but times have changed. Not only have Boas' views completely replaced the old "biological racism" of his time and now hold sway throughout academia, but multiculturalism is also the dominant worldview of most modern democratic societies. White supremacy and its racist cohort are permanently discredited.

So, perhaps it is fitting that today we can turn to the same wellspring of transformative thought as a touchstone for examining not just our relationship with each other as humans, but our species' relationship to the world in which we live and to the animals who inhabit it. We would do well to learn from the people who themselves have gleaned real wisdom from being in the world of whales.

The cornerstone of Kwakwaka'waka religious thought is the codependency of all of nature; no part of the natural order can exist without the rest. There is no such thing as self-sufficiency, whether for humans or their tribes, for animals or the supernatural beings whose powers they represent. Humans are somewhat naturally at the center of their universe, but they accept that all other members of their common world possess not just an indestructible and unique quality, but a spiritual and material parity in that world. "Kwakiutl religion represents the concern of the people to occupy their own proper place within the total system of life, and to act responsibly within it, so as to acquire and control the powers that sustain life," explained Boas' student, Irving Goldman, in his study of the tribe's theology, *The Mouth of Heaven*.

These concerns find their clearest expression in the mythology of animals and the supernatural beings who take their forms. In the Kwak-waka'waka world, humans and animals have real kinship, reflected in the view of killer whales as their ancestors; they have social and spiritual ties that can never be severed. Indeed, they believe that when the tribesmen who hunt marine mammals die, they return to the undersea village of their orca ancestors. In this universe, humans are the recipients of powers, and the givers of those powers are the animals and the supernatural forces they represent. Of all the animals in their universe, the orca is the most powerful, one of the few (along with the raven, the otter, and the wolf) capable of giving a man enough power to become a shaman.

Acquiring a worldview like this does not require us to submit to a belief in supernatural beings, but it does require us to abjure our arrogance, which, as we have seen, is already at the core of our relationship not just with killer whales, but our world generally. Killer whales inherently challenge our assumptions of species superiority, as well as supremacy. Beyond being merely physically more powerful (at least, without tools or technology), orcas can challenge us in the realm of intellectual prowess as well, particularly given the added dimension with which they can gather information about our world and their proven ability to manipulate acoustics to do that. It is also hard to argue with six million years of actual supremacy as the undisputed lords of the oceans when it comes to evolutionary success, species-wise.

Before about 1990, we could reasonably plead ignorance about the unflattering realities that orcas present in relation to humans, especially the way in which what we have learned about them shines a spotlight on our own cognitive limitations. The dirty truth of dolphin and orca studies is that they have established fairly clearly that human beings may well lack the cognitive capacity to understand how all cetaceans communicate; we're just not that acoustically sophisticated.

When we are forced to concede, as with orcas, that we are not unique in our intelligence, that we may not be the only creatures worthy of being considered persons, then we likewise have to reconsider our previous, Western-grown position as special beings somehow separated from nature, with such separation being something desirable instead of the abomination that it would be to someone from the Kwak-waka'waka tribe. It is this latter worldview, one that places humans on an equal, and utterly codependent, footing with nature, as well as the spiritual components that accompany that worldview, that in the cold light of day makes logical sense, especially when we are confronted by the majestic truth that is an orca in full breach or a tall black fin approaching our kayak in the fog.

This realization affects our relationship not just with killer whales, but with all the natural world and with all the animals with whom we share it. It demands that we discard the invented notion of animals as property and recognize that granting them rights does not force us to lose control of the animals we already control; it just requires us to treat them decently.

It also forces us to recognize that we cannot continue degrading and gradually destroying the natural environment that created this bounty of wondrous life, because we are connected to it as deeply as are the wildlife who inhabit it. Our survival as a species, as human beings, of everything that defines us as human, depends on its survival, and so far, it is not looking good for any of us.

In many ways, the attempt to preserve the last remaining killer whales in the Salish Sea is a kind of grand experiment. Never before, for such an extended period, have so many wild orcas lived in close proximity with a massive population of several million human beings. It has never worked well for killer whales in the past; the history of human-orca interaction is almost entirely littered with the corpses of dead cetaceans. So it is a noble effort, even if perhaps doomed, an attempt to prove that human history does not have to lead to the inevitable extinction of every wild thing with which we come in contact, to the destruction of every wild place, every last spark of what makes us human. Only time will tell if it will succeed, but as long as the orcas are still with us, then perhaps we can tell ourselves that there is still hope for us, too.

Recovering our humanity may be the real gift of the orcas, what they can teach us. It's our choice whether to listen.

Acknowledgments

THIS BOOK WAS WRITTEN WITH MATERIAL GATHERED OVER SOME 25 years of observing killer whales and writing about the challenges they face. It was made possible by the openness and willingness of the many scientists who study the whales to spend time with me talking and explaining, and by the many whale activists and wildlife naturalists who take the time to share their knowledge with laymen like myself.

Howard Garrett and Susan Berta of the Orca Network, who have been tireless in their efforts to promote the interests of the Southern Resident killer whales and the salmon that feed them—not to mention to earn the release of Lolita from her tank in Miami—earn special mention. One could not ask for more helpful and generous people to undertake this cause and to guide hapless journalists in the right direction.

I'm also particularly grateful to Ken Balcomb for his many hours spent in interview, as well as his phenomenal crew at the Center for Whale Research, notably Dave Ellifrit and Erin Heydenreich, as well as Astrid Van Ginneken.

A number of other scientists have proved generous with their time as well, particularly Paul Spong and Helena Symonds of British Columbia's OrcaLab. I also would like to thank John K.B. Ford of Canada's Department of Fisheries, Brad Hanson and Lynne Barre of the National Marine Fisheries Service, Lori Marino of the Emory Center for Ethics, Naomi Rose of the Animal Welfare Institute, former UW researcher David Bain, Val Veirs and Scott Veirs of BeamReach, Fred Felleman of Friends of the Earth, and UW researcher Dave Haas, for the time and expertise they've been willing to share.

There are also a number of scientists researching orcas around the world whose work has also proved invaluable in putting together this book, even though I was not able to interview them, notably Craig Matkin and Eva Saulitis, the leading experts on Alaska's killer-whale population; Robin Baird, who has led a number of orca-research projects in the Pacific Northwest for many years, currently as part of Cascadia Research Collective; Lance Barrett-Lennard, one of the Northwest's leading orca scientists; Ingrid Visser, the chief orca scientist working in New Zealand; Phillip Morin of NMFS, one of the leading specialists in orca genetics; and Andrew Foote, whose work with North Atlantic killer whales is proving vital across several fronts.

Bob Otis of Wisconsin's Ripon College, who heads up the summer research team that mans the observation station inside of the lighthouse at Lime Kiln Point State Park, deserves special recognition for his many hours not only observing killer whales there, but also providing an interpretive service for the visitors who want to know more about orcas. Bob was generous with his time in helping guide this book.

We all owe a debt of gratitude to the Whale Museum of Friday Harbor, which has long been one of the main centers of education and activism on behalf of the killer whales and other marine life in the San Juans, and to the many people who have contributed to its ongoing efforts over the years, including former director Rich Osborne and Kari Koski, the longtime director of its SoundWatch program. Special thanks to Jenny Atkinson, the museum's current director, and Cindy Hansen, its education coordinator, for their tireless work.

I'd also like to thank the crew of rangers at San Juan County Park, where I have camped so many times over the years that I nearly qualify for permanent residency, for their superb care and stewardship, as well as their friendship, especially Joe Luma and Eugene Pasinski, as well as the late Ron Abbott.

I also have a debt of gratitude to the people who made it possible for this book to come about, notably my agent, Jill Marsal of Marsal Lyons Agency, who continues to provide smart and wise advice, and who demonstrated remarkable skill in navigating the shift in subject matter this book represents for me, and to my editors at The Overlook Press, Mark Krotov and Dan Crissman, who helped shape the text and make it more accessible. Many thanks as well to my act indexer, Beth Naumann-Montana, and to Allyson Rudolph and Kait Heacock at The Overlook Press for their superb work getting the book across the finish line.

Last of all, I want to thank my daughter, Fiona, for being my inspiration and muse and the reason I write with hope for our future.

Seeing Wild Orcas: A Note

MOST OF THE WORLD'S ORCA POPULATIONS TEND TO OCCUPY waters that are difficult to reach, but North America's Pacific Northwest offers people a chance to see them in the wild for themselves, particularly in two prime locations: The San Juan Islands, and Vancouver Island's interior northern coast.

The San Juan Islands are reachable by ferry from Anacortes, a city about 90 minutes' drive north of Seattle. The ferry ride to San Juan Island, where most of the whale-watching activity occurs, takes about an hour. Friday Harbor has full accommodations for visitors, as does Eastsound on Orcas Island (another whale-watch departure point).

In addition to the boat-based tours, discussed in Chapter Seven, that leave frequently from Friday Harbor, the San Juans offer a nearly unique opportunity to view the whales from land, particularly along the western coast of the main San Juan Island. Land-based whale watching has by far the least impact on the whales, and can often offer you the closest look, particularly at the two most popular spots: Lime Kiln Point State Park, and the County Land Bank, a stretch of open space to the south of the park. In both locations, the orcas are known to frequently come in close to a shoreline that features a 1,300-foot cliff underwater. They can also be viewed, though slightly farther out, with great regularity at San Juan County Park, two miles north of Lime Kiln.

This park is also the main departure point for most of the island's guided kayak tours, and these tours are in the heart of orca-viewing territory. Individual kayakers must obtain a permit—which includes a session on wildlife protocols—from the park rangers before launching.

Northern Vancouver Island is more difficult to reach, but there is a significantly greater orca population there (there are over 200 Northern Residents, while Southern Resident numbers are down to 79) and the setting is far more wild and pristine. You can get there by a couple of driving routes: From Victoria, at the island's southern end (which itself can be reached by ferry either from Anacortes or Port Angeles), you can hit the island's main north-south highway, from which a four-hour drive will get you where you need to be. You can also drive to Vancouver, B.C., and take a ferry to Nanaimo, which will also put you on the same route.

Two towns there provide whale-watching access: The larger city of Port McNeill, which has full accommodations and ferry service to the nearby interior islands, and the tiny, quaint village on tidal stilts that is Telegraph Cove, where there are also full but more limited accommodations, and where most of the kayak tours depart. Individual kayakers also will want to depart there and take the coastal route south to Kaikash Creek and other wilderness camping sites. Both towns have full powerboat-based whale-tour operations that are first-rate. (The nearby island town of Alert Bay, home of the Kwakwaka'wakw tribe and worth visiting just to tour the cultural center and museum, also offers both boat-based tours and kayaking departure points as well. It can be reached by the ferry from Port McNeill.)

Regardless of what kind of craft you use, always remember when you are on the water to respect the orcas, and always keep in mind that these creatures are in a fight for survival and that your presence can stress them and harm them in subtle ways that eventually cause them to lose that fight. In any kind of boat, always maintain the minimum safe distance of 100 yards. Never, ever, position yourself directly in the path of the orcas. If you are in a powerboat, turn off your engines and maintain silence after establishing your position. If you are in a kayak, pull into a cove or get into a kelp bed—at a bare minimum, raft up with other kayakers so that the orcas don't have to run an obstacle course. Even though kayaks don't disrupt with noise, they can still stress orcas by herding them and forcing them to change their course of direction or interfere with their hunting.

Just keep in mind, in all events, that the whales are not there to enthrall humans, much as they might do so. Rather, we are guests in their home, and being respectful and careful is what good guests do.

Epigraphs

pg. ix   "A dolphin appears to be a 'who'": Thomas W. White, *In Defense of Dolphins: The New Moral Frontier* (London: John Wiley & Sons, 2008), p. 184.

"A human being is part of a whole": Albert Einstein, quoted in David Suzuki, *The Sacred Balance: Restoring Our Place in Nature* (Vancouver: Greystone Books, 1999), p. 26.

Chapter 1

pg. 5   I had read her a book: Paul Owen Lewis, *Davy's Dream: A Young Boy's Adventures with Wild Orcas* (Berkeley: Tricycle Press, 1999).

pg. 8   There's a local legend: The story appears in Peter J. Fromm, *Whale Tales: Human Interactions With Whales, Volume Two* (Friday Harbor: Whale Tales Press, 2000), pp. 9–10. The eyewitness account was written by local resident Bruce Conway.

Chapter 2

pg. 16   "Our people have a great respect for the whales: Paul Kennedy, "Legends of the Kwakwaka'wakw," CBC Radio, June 28, 2013. http://www.cbc.ca/ideas/episodes/2013/06/28/legends-of-the-kwakwakawakw/

pg. 16   One legend tells: Various versions of this legend appear in Haida storyteller texts, but the best-known version of it is in Paul Owen Lewis's book for children, *Storm Boy* (Berkeley: Tricycle Press, 2001).

pg. 17   In ancient times: Kennedy, "Legends of the Kwakwaka'wakw."

pg. 18   One of these lives in the San Juan Islands: The book is by Mary J. Getten, *Communicating With Orcas: The Whales' Perspective* (Victoria: Trafford Publishing, 2002).

pg. 20   One of the better known examples: Doug Esser, "Orcas Circle Ferry Transporting Tribal Artifacts to Bainbridge Island," *Seattle Times*, Oct. 31, 2013. http://seattletimes.com/html /localnews /2022166728_orcaferryxml.html

pg. 21   It's true of captive whales: Alexandra Morton, *Listening to Whales: What the Orcas Have Taught Us* (New York: Ballantine Books, 2002), pp. 96–98.

pg. 22   There already has been a considerable backlash: Justin Gregg, *Are Dolphins Really Smart? The mammal behind the myth* (Oxford: Oxford University Press, 2012).

pg. 26   There are plenty of skeptics: Keith Cooper, "Dolphin Studies Could Reveal Secrets of Extraterrestrial Intelligence," *Astrobiology Magazine*, Sept. 2, 2011.

pg. 28   Dolphin scientist Thomas White: Thomas I. White, *In Defense of Dolphins: The New Moral Frontier* (Oxford: Blackwell Publishing, 2007).

pg. 31   Lori Marino observes: For more details on Marino's findings on the anatomy of the orca brain, see Lori Marino, Chet C. Sherwood, Bradley N. Delman, Ceuk Y. Tang, Thomas P. Nadich, and Patrick R. Hof, "Neuroanatomy of the Killer Whale (*Orcinus orca*) From Magnetic Resonance Images," *The Anatomical Record*, Part A (2004), 1256–1263.

Chapter 3

pg. 39   That describes the ethereal daily life: Two excellent explorations of the daily life of killer whales can be found in Douglas H. Chadwick, *The Grandest of Lives: Eye to Eye With Whales,* (San Francisco: Sierra Club Books, 2006) pp. 119–167, and Astrid Van Ginneken, *Togetherness Is Our Home: An Orca's Journey Through Life* (Friday Harbor: Booksurge, 2007).

pg. 43   The sophistication of dolphin-family echolocation: Tony Perry, "Navy Dolphins Discover Rare Old Torpedo Off Coronado," *Los Angeles Times,* May 17, 2013. http://articles.latimes.com/2013/may/17/local/la-me-torpedo-dolphins-20130518

pg. 43   Orcas' sonic capacities: Louis Bergeron, "'Orca Ears' Inspire Stanford Researchers to Develop Ultrasensitive Undersea Microphone," Stanford University News Service, June 23, 2011. http://news.stanford.edu/pr/2011/pr-orca-ears-microphone-062311.html

pg. 50   There is a constellation: Justin Gregg, *Are Dolphins Really Smart?,* pp. 135–182.

pg. 52   Orca vocalizations typically are: The definitive study of resident orca vocalizations remains John K.B. Ford's "Acoustic Behaviour of Resident Killer Whales Off Vancouver Island, British Columbia," *Canadian Journal of Zoology,* 1989, 67(3): 727–745, 10.1139/z89–105. However, subsequent studies have substantially helped to clarify the picture, notably: Volker Bernt Deeke, "Stability and Change of Killer Whale (*Orcinus Orca*) Dialects," University of British Columbia, June 1988; Rüdiger Riesch, John K.B. Ford, Frank Thomsen, "Stability and group specificity of stereotyped whistles in resident killer whales, Orcinus orca, off British Columbia," *Animal Behaviour,* Volume 71, Issue 1, January 2006, pp. 79–91; Olga A. Filatova, Volker B. Deecke, John K.B. Ford, Craig O. Matkin, Lance G. Barrett-Lennard, Mikhail A. Guzeev, Alexandr M. Burdin, and Erich Hoyt, "Call diversity in the North Pacific killer whale populations: implications for dialect evolution and population history," *Animal Be-*

haviour, Volume 83, Issue 3, March 2012, pp. 595–603; Volker B. Deeke, John K.B Forde, and Paul Spong, "Dialect change in resident killer whales: implications for vocal learning and cultural transmission,"*Animal Behaviour*, Volume 60, Issue 5, November 2000, pp. 629–638; Richard Riesch and V.B. Deecke, "Whistle communication in mammal-eating killer whales (Orcinus orca): further evidence for acoustic divergence between ecotypes," *Behavioral Ecology and Sociobiology* 65:1377–1387, 2011; and Nicola Rehn, Stefanie Teichert, and Frank Thomsen, "Structural and Temporal Emission Patterns of Variable Pulsed Calls in Free-Ranging Killer Whales," *Behaviour*, 144, 307–329, 2007.

pg. 57   A team of researchers: The finished study is Andrew D. Foote, Rachael M. Griffin, David Howitt, Lisa Larsson, Patrick J.O. Miller, and A. Rus Hoelzel, "Killer Whales Are Capable of Vocal Learning," *Biology Letters*, doi: 10.1098 /rsbl.2006.0525.

pg. 58   That was the adjective: See the documentary, *The Whale* (2011).

pg. 58   Two journalists who arrived: See Michael Parfit and Suzanne Chisholm, *The Lost Whale: The True Story of an Orca Named Luna* (New York: St. Martins Press, 2013). A video of Luna imitating a boat motor can be seen at http://www.youtube .com/watch?v=3X8nIXTtgBk.

Chapter 4

pg. 68   There are some variations: The main version of this myth can be found at Mary L. Beck, *Heroes and Heroines in Tlingit-Haida Legend*, (Seattle: Alaska Northwest Books, 1989), pp. 3–14. One of the variations can be found at Ghandl of the Qayahl Llaanas (translated by Robert Bringhurst), *Nine Visits to the Mythworld* (Vancouver: Douglas & McIntyre, 2000), pp. 97–110. The version involving the sea otter can be found at Wikipedia: http://en.wikipedia.org/wiki/Natsilane

pg. 69   A Southeast-Alaskan Tsimshian myth: This is from Shannon Thunderbird's online collection of Northwest Native oral traditions, "Animal Stories and Legends and Teachings: Orca ('Neexl)," http://www.shannonthunderbird.com/stories%20n-z.htm.

pg. 70   One of the land ancestors of all whales: See Annalisa Berta, James L. Sumich, and Kit M. Kovacs, eds., *Marine Mammals: Evolutionary Biology* (New York: Academic Press, 2006), pp. 57–81

pg. 71   The fossils of the *Basilosaurus*: See Herman Melville, *Moby Dick, or The Whale*, Chapter CIV, "The Fossil Whale."

pg. 72   *Delphinidae* are the largest: See William F. Perrin, Bernd Wursig, and J.G.M. Thewissen, eds., *Encyclopedia of Marine Mammals* (London: Academic Press, 2009), pp. 298–302.

pg. 73   It was Bigg who had pioneered: See Bruce Obee and Graeme Ellis, *Guardians of the Whales: The Quest to Study Whales in the Wild* (Anchorage: Alaska Northwest Books, 1992), pp. xi–xiii, and Erich Hoyt, *Orca: The Whale Called Killer* (Ontario: Camden House, 1990), pp. 69–108.

pg. 75   It was somewhat unprecedented. See Associated Press, "Orcas Devouring Harbor Seals," Feb. 24, 2003, http://www.seattlepi.com/news/article/Orcas-devouring-harbor-seals-1108256.php

pg. 75   Dubbed "the slippery six": See Associated Press, "Six Killer Whales Extend Stay in Puget Sound for Record Time," July 29, 2005, http://articles.latimes.com/2005/jul/09/nation/na-orcas9.

pg. 76   Finally, in 2003: See K.M. Parsons, J.W. Durban, D. E. Claridge, "Comparing two alternative methods for genetic sampling of small cetaceans," *Marine Mammal Science*, 19:224–231. An earlier study—Lance Barrett-Lennard, "Population Structure and Patterns of Killer Whales (*Orcinus orca*) as Revealed by DNA Analysis," University of British Columbia, December 2000—indicated similar results. See also P.A. Morin, R.G. LeDuc, K.M. Robertson, N.M. Hedrick, W.F. Perrin, M. Etnier, P. Wade, B.L. Taylor, "Genetic Analysis of Killer Whale

(*Orcinus orca*) Historical Bone and Tooth Samples to Identify Western U.S. ecotypes," *Marine Mammal Science* 22, 897–909, 2006, and Kim M. Parsons, John W. Durban, Alexander M. Burdin, Vladimir N. Burkanov, Robert L. Pitman, Jay Barlow, Lance G. Barrett-Lennard, Richard G. LeDuc, Kelly M. Robertson, Craig O. Matkin and Paul R. Wade, "Geographic Patterns of Genetic Differentiation among Killer Whales in the Northern North Pacific," *Journal of Heredity*, Volume 104, Issue 6, pp. 737–754, 2013.

pg. 77   By 2011, a general consensus: See Lance Barrett-Lennard, "Killer Whale Evolution: Populations, Ecotypes, Species, Oh, My!", *Journal of the American Cetacean Society*, Vo. 40, No. 1, pp. 48–53.

pg. 77   The debate focused around: See P.A. Morin, F.I. Archer, A.D. Foote, J. Vilstrup, E.E. Allen, P.R. Wade, J.W. Durban, K.M. Parsons, R. Pitman, L. Li, et al., "Complete Mitochondrial Genome Phylogeographic Analysis of Killer Whales (*Orcinus orca*) Indicates Multiple Species," *Genome Research*, 20:908–91, 2010.

pg. 84   In other orca ecotypes: See Robin Baird and Hal Whitehead, "Social Organization of Mammal-Eating Killer Whales: Group Stability and Dispersal Patterns" *Canadian Journal of Zoology*, 78: 2096–2105 2000, as well as Robin W. Baird and Lawrence M. Dill, "Ecological and Social Determinants of Group Size in Transient Killer Whales," *Behavioral Ecology*, Volume 7, Issue 4, pp. 408–416.

pg. 85   The best-documented case: See John K.B. Ford and Graeme M. Ellis, *Transients: Mammal-Hunting Killer Whales* (Seattle: University of Washington Press, 1999), pp. 20–21.

pg. 89   Sting-Ray Teamwork: See Ingrid Visser, et. al, "Benthic Foraging on Stingrays by Killer Whales (*Orcinus Orca)* in New Zealand Waters," *Marine Mammal Science*, 15(1): 220–227, January 1999.

pg. 90   Beach Snatchers: A video of a killer whale snatching a baby

elephant seal at Valdes can be viewed at https://www.youtube
.com/watch?v=N53DchkdUUM

pg. 90   Ice-Floe Harmonics: A video of a team of killer whales using harmonics to force a seal from an ice floe can be viewed at https://www.youtube.com/watch?v=p3xmqbNsRSk

pg. 91   As scientists: See Luke Rendell and Hal Whitehead, "Culture in Whales and Dolphins," *Behavioral and Brain Sciences*, 24 (2):309–324, 2001.

Chapter 5

pg. 95   Four sisters are out walking: This story is one of several oral-history legends that can be heard online at "Legends of the Kwakwaka'wakw," CBC Radio, June 28, 2013. http://www.cbc.ca /ideas/episodes/2013/06/28/legends-of-the-kwakwakawakw/

pg. 97   In the telling of the S'Klallam tribe: See "Kakantu, the Chief's Daughter Who Married a Blackfish: A Traditional Klallam Story," told by Amy Allen, translated by Lawrence C. Thompson and Martha Charles John, in M. Terry Thompson, and Steven M. Egesdal, *Salish Myths and Legends: One People's Stories* (Lincoln: University of Nebraska Press, 2008), pp. 401–404.

pg. 98   There were other cultures: See Hans Rollmann, "Religion in Newfoundland and Labrador," Newfoundland and Labrador Heritage, Memorial University of Newfoundland, http://www .heritage.nf.ca/society/religion.html. See also the Wikipedia entry on Shachihoko, http://en.wikipedia.org/wiki/Shachihoko

pg. 98   The Japanese, dating back to at least: See Toshio Kasuya, "Japanese Whaling," in Perrin, William F., Wursig, Bernd, and Thewissen, J.G.M., eds. *Encyclopedia of Marine Mammals* (London: Academic Press, 2009), pp. 643–649.

pg. 98   In Siberia, the indigenous Yupik people: See "The Orphan Boy With His Sister," p. 156 in E. S. Rubcova, *Materials on the Language and Folklore of the Eskimoes, Vol. I* (Chaplino Dialect. Leningrad: Academy of Sciences of the USSR), 1954.

pg. 99   They frightened Pliny the Elder: See Gaius Plinius Secundus, *Historia Naturalis* 9.5.12 (Latin).

pg. 100   According to Ariosto: See Lodovico Ariosto, *The Orlando Forioso, Vol. 1*, translated by William Stewart Rose (London: Henry G. Bohn, 1854), pp. 173–177. Online at http://books.google.com/books?id=LlkJAAAAQAAJ&pg=PA173&lpg=PA173

pg. 101   These same Basques: See Eric Jay Dolin, *Leviathan: The History of Whaling in America* (New York: W.W. Norton & Co., 2001), pp. 22–23.

pg. 101   There was one whaling operation, however: See the documentary by Klaus Toft, Klaus, *Killers in Eden* (Australian Broadcasting Corporation, 2007), online at http://www.pbs.org/wnet/nature/episodes/killers-in-eden/introduction/1048/. A website is also devoted to the subject, titled "Killers of Eden," http:/ /www.killersofeden.com/

pg. 103   "Three or four of these voracious animals: Scammon quoted in Obee and Ellis, *Guardians of the Whales*, p. 6.

pg. 104   Earlier that same summer: See Dougal Robertson, *Survive the Savage Sea* (Dobbs Ferry, N.Y.: Sheridan House, 1994).

pg. 105   There was, as it happens: See Nathaniel Philbrick, *In the Heart of the Sea: The Tragedy of the Whaleship Essex* (New York: Penguin Books, 2001), pp. 115–116.

pg. 105   A 1963 book: See Joseph J. Cook, *Killer Whale!* (New York: Dodd, Mead, 1963).

pg. 106   Then, in November of 1961: Brocato was interviewed by *Frontline* for its 1997 investigative documentary report, *A Whale of a Business*. Online at http://www.pbs.org/wgbh/pages/frontline/shows/whales/

pg. 107   Finally, in 1964: See Hoyt, *Orca: The Whale Called Killer*, pp. 16–19, 113–26, 238–52. See also Daniel Francis and Gil Hewlett, *Operation Orca: Springer, Luna and the Struggle to Save West Coast Killer Whales* (Vancouver: Harbor Publishing, 2007), pp. 60–66.

pg. 109 Tors was in the middle: Ted Griffin was also interviewed for PBS for its documentary *A Whale of a Business*, online at http://www.pbs.org/wgbh/pages/frontline/shows/whales/ .

pg. 114 Apparently Shamu had been: David Kirby, *Death at SeaWorld: Shamu and the Dark Side of Killer Whales in Captivity* (New York: St. Martins Press, 2012), pp. 168, 173–174.

pg. 118 One of the worst of these: See Sandra Pollard, *Puget Sound Whales for Sale: The Fight to End Orca Hunting* (Charleston, S.C.: The History Press, 2014), pp. 74–90.

See also the website devoted to the incident, "The Penn Cove Captures," http://us .whales.org/issues/penn-cove-orca -captures

pg. 119 The issue came to a head: See Pollard, pp. 133–151.

pg. 122 The first effort: See Hoyt, *Orca: The Whale Called Killer*, pp. 114–127.

pg. 123 His trainer, a young Canadian: See Vivian Kuo, "Orca Trainer Saw Best of Keiko, Worst of Tilikum," CNN, Oct. 28, 2013. Online at http://www.cnn.com/2013/10/26/world/americas/orca -trainer-tilikum-keiko/

See also Tim Zimmerman, "The Killer in the Pool," *Outside*, July 30, 2010. Online at http://www.outsideonline.com/outdoor -adventure/nature/The-Killer-in-the-pool.html.

pg. 124 In 1999: See Kirby, *Death at SeaWorld*, pp. 257–260.

pg. 125 John Jett: See the documentary *Blackfish* (2013), in which Jett is interviewed.

pg. 126 A few right-wing voices: See Kyle Mantyla, "Fischer: Sea-World Death Due to West's Failure to Follow Scripture," Right Wing Watch, Feb. 25, 2010, http://www.rightwingwatch.org /content/fischer-sea-world-death-due-wests-failure-follow -scripture

pg. 128 During a panel discussion on *Crossfire:* See David Neiwert, "Dodging 'Blackfish': What SeaWorld Doesn't Want You To Know," Crooks and Liars, Oct. 29, 2013, http://crooksandliars

.com/david-neiwert/dodging-blackfish-why-sea-world-does

pg. 128   Rather than back down: The website 'The Truth About Blackfish' is at http://seaworld.com/truth/truth-about-blackfish/ .

pg. 128   However, shortly after the SeaWorld website: See Jason Garcia, "Blackstone Chief Blames Brancheau for Own Death, Contradicting SeaWorld," *Orlando Sentinel*, Jan. 24, 2014. http://www.orlandosentinel.com/business/tourism/tourism-central-florida-blog/os-blackstone-chief-blames-brancheau-for-own-death-contradicting-seaworld-20140124-post.html

Chapter 6

pg. 132   Skana was spunky: See Hoyt, *Orca: The Whale Called Killer*, pp. 41–44. See also Francis and Hewlett, *Operation Orca*, pp. 67–71, and Obee and Ellis, *Guardians of the Whales*, pp. 16–19.

pg. 136   Mike Bigg also showed up: See Hoyt, *Orca: The Whale Called Killer*, pp. 69–83. See also Obee and Ellis, *Guardians of the Whales*, pp. 19–24.

pg. 138   In 1983, he was first: See Obee and Ellis, *Guardians of the Whales*, pp. xii–xiii.

pg. 144   Graeme Ellis knew Bigg, too: See Obee and Ellis, *Guardians of the Whales*, pp. 11–16, 21–27.

pg. 145   Alexandra Morton, like most: Morton tells her own story eloquently in her book *Listening to Whales: What Killer Whales Have Taught Us* (New York: Ballantine Books, 2002).

Chapter 7

pg. 164   In Alaska, the resident killer whale population: See especially Eva Saulitis, *Into Great Silence: A Memoir of Discovery and Loss Among Vanishing Orcas*, (Boston: Beacon Press, 2013).

pg. 166   Salmon are a uniquely useful species: See David R. Montgomery, *King of Fish: The Thousand-Year Run of Salmon* (Boulder: Westview Press, 2003).

pg. 168   That was nothing, of course: See Jim Lichatowich, *Salmon Without Rivers: A History of the Pacific Salmon Crisis* (Washington, D.C.: Island Press, 1999), and Bruce Brown, *Mountain in the Clouds: A Search for the Wild Salmon* (Seattle: University of Washington Press, 1995)

pg. 170   These arguments, however: See Mike Lee and Kim Bradford, "Thousands Rally to Save Snake Dams," *Tri-City Herald*, Feb. 19, 1999.

pg. 171   The decision to list the orcas: See "Endangered Species Status of Puget Sound Killer Whales," NOAA Fisheries West Coast Region, 2014. http://www.westcoast.fisheries.noaa .gov /protected_species/marine_mammals/killer_whale/esa _status.html

pg. 171   Probably the best public demonstration: See Jim Simon, "Orcas Put Bite On Salmon Catch—Pod's Appetite Prompts Closure Of Dyes Inlet Fishery," *Seattle Times*, Nov. 13, 1997.

pg. 173   Once that was decided: See Morin, et. al., "Complete Mitochondrial Genome Phylogeographic Analysis of Killer Whales (*Orcinus orca*) Indicates Multiple Species," *Genome Research*, 20:908–91, 2010.

pg. 179   This is all part of a grand experiment: See the documentary *Undamming the Elwha*, produced by Katie Campbell and Michael Werner for KCTS-TV, http://kcts9.org/undamming-elwha and the documentary *Return of the River*, by John Gussman and Jessica Plumb, http://www.elwhafilm.com/synopsis.htm ,

pg. 182   It took some work: See Robert McClure, "Dead orca is a 'red alert,'" *Seattle Post-Intelligencer*, May 7, 2002, http://www.eurocbc .org/page96.html

pg. 183   There are three specific kinds of POPs: See D.L. Cullon, M.B. Yunker, C. Alleyne, N.J. Dangerfield, S. O'Neill, M.J. Whiticar, P.S. Ross, "Persistent Organic Pollutant in Chinook Salmon (Oncorhynchus tshawytscha): Implications for Resident Killer Whales of British Columbia and Adjacent Waters," *Environmental Toxicology and Chemistry* 28(1): 148–161, 2009.

See also Cathy Britt, "The Killer Affecting Killer Whale Populations," *QUEST Northwest*, July 19, 2011, http://science .kqed.org/quest/2011/07/19/the-killer-affecting-killer-whale -populations/

pg. 187 Scientists have been studying: See David E. Bain, Jodi C. Smith, Rob Williams, and David Lusseau, "Effects of Vessels on Behavior of Southern Resident Killer Whales," NMFS Contract Report, March 4, 2006, and Rob Williams, Andrew W. Trites, and David E. Bain, "Behavioural Responses of Killer Whales (*Orcinus Orca*) to Whale-Watching Boats: Opportunistic Observations and Experimental Approaches," *Journal of the Zoological Society of London*, 256, 255–270, 2002.

pg. 191 Research released in the summer of 2014: See "Causes of Decline Among Southern Resident Killer Whales," Center for Conservation Biology, University of Washington, June 2014, http://conservationbiology.uw.edu/research-programs/killer -whales/. See also Ayres, Katherine L., Rebecca K. Booth, Jennifer A. Hempelmann, Kari L. Koski, Candice K. Emmons, Robin W. Baird, Kelley Balcomb-Bartok, M. Bradley Hanson, Michael J. Ford, Samuel K. Wasser, "Distinguishing the Impacts of Inadequate Prey and Vessel Traffic on an Endangered Killer Whale (*Orcinus orca*) Population," PLoS ONE 7(6): e36842, 2012.

Chapter 8

pg. 212 In 2013, Russian fishermen: See Erich Hoyt, "Russian Orca Captures: The Inside Story," *Whale and Dolphin Conservation*, Nov. 11, 2013. Online at http://us.whales.org/blog/erichhoyt /2013/11/russian-orca-captures-inside-story

pg. 215 When PBS's *Frontline*: See the transcript at http://www.pbs.org /wgbh/pages/frontline/shows/whales/interviews/mccaw1.html

pg. 215 Aquarium director Phyllis Bell: See transcript at http://www .pbs.org/wgbh/pages/frontline/shows/whales/etc/script.html

pg. 215 Phyllis Bell was so enraged: See Kirby, *Death at SeaWorld*, pp. 243–244. See also Associated Press, "Executive Director at

Oregon Coast Aquarium Resigns," July 4, 2002, http://www
.seattlepi.com/news/article/Executive-director-at-Oregon
-Coast-Aquarium-1090674.php, and Larry Bacon, "Dealings of
Bell Will Be Investigated," *Eugene Register Guard*, July 26, 2002,
http://news.google.com/newspapers?nid=1310&dat=20020726
&id=O75YAAAAIBAJ&sjid=1OsDAAAAIBAJ&pg=2503,5960930,
and "Ex-Aquarium Chief Guilty of Forgery," *Lincoln City News
Guard*, Oct 25, 2003.

pg. 216 When Rose paid a subsequent visit: See Kirby, *Death at Sea-
World*, pp. 266–270.

pg. 218 "We were sort of waiting: See the documentary, *Keiko: The
Untold Story* (2010), which features interviews with Baird.
http://www.keikotheuntoldstory.com/

pg. 220 "My belief is that Keiko: See Paul Spong, "Keiko's Incredible
Journey," Earth Island Institute, March 19, 2010. http://keiko
.com/pSpongStatement.html

pg. 221 Springer was first noticed: See Leigh Calvez, "Springer the Lost
Orca Part 2: Orca Behavior and a Whale of a Mystery Solved,"
Inside Bainbridge, July 24, 2013. http://www.insidebainbridge
.com/tag/mark-sears/

　　　See also Francis and Hewlett, *Operation Orca*, pp. 87–169.

pg. 224 Springer has not only been: See Keven Drews, "Springer, the
rescued orphaned killer whale, spotted with calf off Vancouver
Island," *The Canadian Press*, July 9, 2013, http://www.thestar.com
/news/canada/2013/07/09/springer_the_rescued_orphaned
_killer_whale_spotted_with_calf_off_vancouver_island.html

pg. 230 Most of all, we now know: See Naomi Rose, "Killer Controversy: Why Orcas Should No Longer Be Kept in Captivity,"
Humane Society International, 2011. http://www.hsi.org/assets
/pdfs/orca_white_paper.pdf

pg. 232 SeaWorld stock: See David Neiwert, "SeaWorld and Its Terrible, Horrible, No Good, Very Bad Week," Crooks and Liars,
Aug. 18, 2014, http://crooksandliars.com/2014/08/seaworld
-and-its-terrible-horrible-no-good

pg. 236 In January 2014: See Craig Welch, "Feds Want Endangered Status for Captive Orca Lolita," *Seattle Times,* Jan. 24, 2014, http://seattletimes.com/html/localnews/2022749958_lolitaxml.html

pg. 237 Lolita is also at the center: See Sasha Luque, "Animal Activists File Lawsuit to Free Lolita," NBC Miami, Nov. 18, 2011, http://www.nbcmiami.com/news/Animal-Activists-file-lawsuit-to-try-to-free-Lolita-134121753.html

pg. 241 Damning video footage: The footage can be viewed at https://www.youtube.com/watch?v=wt2F9iYCff8&feature=youtu.be

See also Madeline Bar-Diaz, "Aquatic Attraction Cited for Code Violations," *Sun Sentinel,* Sept. 13, 2003, http://articles.sun-sentinel.com/2003-09-13/news/0309130059_1_animals-or-employees-code-violations-miami-dade-county-building

and Andres Viglucci, "Miami Seaquarium Allowed to Open After Most-Urgent Repairs Completed," *Miami Herald,* Sept. 16, 2003, http://www.highbeam.com/doc/1G1-107815075.html

and Associated Press, "Repaired Seaquarium Reopens Today," *Tampa Bay Times,* Feb. 11, 2006, http://www.sptimes.com/2006/02/11/State/Repaired_Seaquarium_r.shtml

pg. 242 A twist in Lolita's story: See Hannah Sampson, "California Theme Park Company to Buy Miami Seaquarium," *Miami Herald,* March 28, 2014.

pg. 244 In 2010, Robert Rose: See Robert Samuels, "Lolita Still Thrives at Miami Seaquarium," McClatchy News Service, Sept. 15, 2010. http://seattletimes.com/html/travel/2012903643_webwhale16.html

Chapter 9

pg. 247 Three Kwakwaka'wakw boys: This story is one of several oral-history legends that can be heard online at "Legends of the Kwakwaka'wakw," CBC Radio, June 28, 2013. http://www.cbc.ca/ideas/episodes/2013/06/28/legends-of-the-kwakwakawakw/

pg. 250   "We ignore the inconvenient fact: See Jeff Schweitzer, "The Dirty Little Secret About Human Intelligence," Huffington Post, Dec. 15, 2013, http://www.huffingtonpost .com/jeff-schweitzer/social-hour-understanding_b_4450076 .html

pg. 250   Because personhood has always: See Thomas I. White, *In Defense of Dolphins: The New Moral Frontier* (Oxford: Blackwell Publishing, 2007), p. 152.

pg. 251   Brain specialist Harry Jerison: See Harry J. Jerison, "The Perceptual World of Dolphins," in Ronald J. Schusterman, Jeanette A. Thomas, and Forrest G. Wood, eds., *Dolphin Cognition and Behavior: A Comparative Approach* (Hillsdale, N.J.: Lawrence Erlbaum Associates, 1986), pp. 141–164.

pg. 254   "No one has ever demanded: See Press Release re: NhRP Lawsuit, Dec. 2nd 2013, Nonhuman Rights Project, http://www .nonhumanrightsproject.org/2013/11/30/press-release-re-nhrp -lawsuit-dec-2nd-2013/

pg. 256   Naomi Rose, herself a onetime: See Naomi Rose, "A Win-Win Solution for Captive Orcas and Marine Parks," CNN.com, Oct. 28, 2013. http://www.cnn.com/2013/10/24/opinion/blackfish -captive-orcas-solutions/

pg. 260   Franz Boas, the father of modern anthropology: See Franz Boas, *Indian Myths and Legends from the North Pacific Coast of America* (Dietrich Bertz, translator), Vancouver: Talonbooks, 2002. See also Franz Boas, "The Social Organization of the Kwakiutl," *American Anthropologist*, Vol. 2, No.2, April-June 1920.

pg. 262   The cornerstone of Kwakwaka'waka religious thought: See Irving Goldman, *The Mouth of Heaven: An Introduction to Kwakiutl Religious Thought* (New York: John Wiley & Sons, 1975), p. 177.

Bibliography

Adamson, Thelma, ed. *Folk-Tales of the Coast Salish*. Lincoln: University of Nebraska Press, 2009.

Baird, Robin W. *Killer Whales of the World: Natural History and Conservation*. Stillwater, MN: Voyageur Press, 2002.

Beck, Mary L. *Heroes and Heroines in Tlingit-Haida Legend*. Seattle: Alaska Northwest Books, 1989.

Berta, Annalisa, Sumich, James L., and Kovacs, Kit M., eds. *Marine Mammals: Evolutionary Biology*. New York: Academic Press, 2006.

Boas, Franz. *Indian Myths and Legends from the North Pacific Coast of America* (Dietrich Bertz, translator). Vancouver: Talonbooks, 2002.

Bringhurst, Robert. *A Story As Sharp As a Knife: The Classical Haida Storytellers and Their World*. Vancouver: Douglas & McIntyre, 2011.

Brown, Bruce. *Mountain in the Clouds: A Search for the Wild Salmon*. Seattle: University of Washington Press, 1995.

Chadwick, Douglas H. *The Grandest of Lives: Eye to Eye With Whales*. San Francisco: Sierra Club Books, 2006.

Ford, John K.B., Ellis, Graeme M., and Balcomb, Kenneth C. *Killer Whales*. Seattle: University of Washington Press, 1994.

Ford, John K.B., and Ellis, Graeme M. *Transients: Mammal-Hunting Killer Whales*. Seattle: University of Washington Press, 1999.

Francis, Daniel, and Hewlett, Gil. *Operation Orca: Springer, Luna and the Struggle to Save West Coast Killer Whales*. Vancouver: Harbor Publishing, 2007.

Fromm, Peter J. *Whale Tales: Human Interactions With Whales (Volumes One and Two)*. Friday Harbor: Whale Tales Press, 1996.

Giraudo Beck, Mary. *Shamans and Kushtakas: North Coast Tales of the Supernatural.* Seattle: Alaska Northwest Press, 1999.

Goldman, Irving. *The Mouth of Heaven: An Introduction to Kwakiutl Religious Thought.* New York: John Wiley & Sons, 1975.

Gregg, Justin. *Are Dolphins Really Smart? The mammal behind the myth.* Oxford: Oxford University Press, 2012.

Hoare, Philip. *The Whales: In Search of the Giants of the Sea.* New York: Ecco, 2010.

Hoyt, Erich. *Orca: The Whale Called Killer.* Ontario: Camden House, 1990.

Kirby, David. *Death at SeaWorld: Shamu and the Dark Side of Killer Whales in Captivity.* New York: St. Martins Press, 2012.

Lichatowich, Jim. *Salmon Without Rivers: A History of the Pacific Salmon Crisis.* Washington, D.C.: Island Press, 1999.

Mann, Janet, Connor, Richard C., Tyack, Peter L., and Whitehead, Hal. *Cetacean Societies: Field Studies of Dolphins and Whales.* Chicago: University of Chicago Press, 2000.

Matkin, Craig O. *The Killer Whales of Prince William Sound.* Valdez: Prince William Sound Books, 1994.

Montgomery, David R. *King of Fish: The Thousand-Year Run of Salmon.* Boulder: Westview Press, 2003.

Morton, Alexandra. *Listening to Whales: What the Orcas Have Taught Us.* New York: Ballantine Books, 2002.

Nollman, Jim. *The Charged Border: Where Whales and Humans Meet.* New York: Henry Holt, 1999.

Obee, Bruce, and Ellis, Graeme. *Guardians of the Whales: The Quest to Study Whales in the Wild.* Anchorage: Alaska Northwest Books, 1992.

Parfit, Michael, and Chisholm, Suzanne. *The Lost Whale: The True Story of an Orca Named Luna.* New York: St. Martins Press, 2013.

Perrin, William F., Wursig, Bernd, and Thewissen, J.G.M., eds. *Encyclopedia of Marine Mammals.* London: Academic Press, 2009.

Pryor, Karen, and Norris, Kenneth, eds. *Dolphin Societies: Discoveries and Puzzles*. Berkeley: University of California Press, 1998.

Reid, Bill, and Bringhurst, Rober. *The Raven Steals the Light*. Seattle: University of Washington Press, 1996.

Robertson, Dougal. *Survive the Savage Sea*. Dobbs Ferry, N.Y.: Sheridan House, 1994.

Rothenberg, David. *The Thousand Mile Song: Whale Music in a Sea of Sound*. New York: Basic Books, 2008.

Saulitis, Eva. *Into Great Silence: A Memoir of Discovery and Loss among Vanishing Orcas*. Boston: Beacon Press, 2013.

Schusterman, Ronald J., Thomas, Jeanette A, and Wood, Forrest G., eds. *Dolphin Cognition and Behavior: A Comparative Approach*. Hillsdale, N.J.: Lawrence Erlbaum Associates, 1986.

Thompson, M. Terry, and Egesdal, Steven M. *Salish Myths and Legends: One People's Stories*. Lincoln: University of Nebraska Press, 2008.

Van Ginneken, Dr. Astrid. *Togetherness is Our Home: An Orca's Journey Through Life*. Friday Harbor: Booksurge, 2007.

Visser, Ingrid. *Swimming with Orca: My Life with New Zealand's Killer Whales*. Auckland: Penguin NZ, 2005.

White, Thomas I. *In Defense of Dolphins: The New Moral Frontier*. Oxford: Blackwell Publishing, 2007.

Index

Note: Page numbers in italic refer to illustrations.